Springer Series in
Nuclear
and **Particle Physics**

Springer Series in **Nuclear** and **Particle Physics**

Editors: Mary K. Gaillard · J. Maxwell Irvine · Vera Lüth · Bruce McKellar · Achim Richter ·

Hasse, R.W., Myers, W.D.
Geometrical Relationships of Macroscopic Nuclear Physics

Belyaev, V.B.
Lectures on the Theory of Few-Body Systems

Heyde, K.L.G.
The Nuclear Shell Model

Gitman, D.M., Tyutin I.V.
Quantization of Fields with Constraints

Sitenko, A.G.
Scattering Theory

Fradkin, E.S., Gitman, D.M., Shvartsman, S.M.
Quantum Electrodynamics with Unstable Vacuum

Brenner, M., Lönnroth, T., Malik, F.B. (Editors)
Clustering Phenomena in Atoms and Nuclei

Makhankov, V.G., Rybakov, Y.P., Sanyuk, V.I.
The Skyrme Model

Antonov, A.N., Hodgson, P.E., Petkov, I.Zh.
Nucleon Correlations in Nuclei

Akhiezer, A.I., Sitenko, A.G., Tartakovskii, V.K.
Nuclear Electrodynamics

Dey, M., Dey, J.
Nuclear and Particle Physics

Mira Dey Jishnu Dey

Nuclear and Particle Physics

The Changing Interface

With 6 Figures

Springer-Verlag
Berlin Heidelberg New York
London Paris Tokyo
Hong Kong Barcelona
Budapest

Dr. Mira Dey
Lady Brabourne College
Calcutta 700017
India

Prof. Jishnu Dey
IFT-UNESP
Rua Pamplona 145
São Paulo 01405-900
Brasil

ISBN-13:978-3-642-84967-1   e-ISBN-13:978-3-642-84965-7
DOI: 10.1007/978-3-642-84965-7

Library of Congress Cataloging-in-Publication Data, Dey, Mira, 1944-Nuclear and particle physics: the changing interface / Mira Dey, Jishnu Dey.   p.   cm. – (Springer series in nuclear and particle physics). Includes
bibliographical references and index. ISBN-13:978-3-642-84967-1
1. Particles (Nuclear physics) I. Dey, Jishnu, 1943-   . II. Title. III. Series. QC793.2.D49 1993
539.7'2 – dc20   93-14273

© Springer-Verlag Berlin Heidelberg 1994
Softcover reprint of the hardcover 1st edition 1994

Typesetting: Data conversion by Laser Words, Madras
56/ 3140/ SPS – 5 4 3 2 1 0 – Printed on acid-free paper

Preface

There is a story of blind men inspecting parts of an elephant and describing it differently by the trunk, ears, legs and tail. The situation is much the same in this book. For example, the proton is sometimes a bag with 3 quarks, sometimes *with other particles too*. Elsewhere it is a topological knot in a meson field, and in another context its mass is a numerical factor multiplying the quark condensate. But this is fascinating, like an elephant!

This book originated from lectures given by one of us at IFT, São Paulo. Profs. Paulo Leal Ferreira and Lauro Tomio encouraged us to write the book. We thank all our collaborators for teaching us many things at various stages. We are also grateful to Drs. Luc Vinet and Avinash Khare for making particle physics look easy from the nuclear physics point of view. Visits to TRIUMF in Vancouver, LINAC in Saskatoon, University of Alberta in Edmonton, ITEP in Moscow and ICTP in Trieste were useful. Drs. Mike Birse and Judith McGovern encouraged us at the later stages of the work through meaningful suggestions. We are inspired by our editor Prof. Max Irvine, whose book on neutron stars contained a mixture of different subjects like ours. He patiently went through our manuscript again and again and suggested interesting changes, all of which we gratefully incorporated. We are happy to get an opportunity to thank our friend Mrs. M. K. Easlea, always so kind to us, for reading the proof thoroughly and pointing out very many important corrections and typographic mistakes.

We have tried to describe work that we know, possibly omitting many contributions from many individuals and groups. We apologize to them. We have tried to give an up-to-date account of developments which we hope will interest the reader the way it interested us.

This book is for Damini, Rukmini, Joyoni and Lynne, the younger generations of our family.

October 1993

Mira Dey
Jishnu Dey

Contents

1   **Introduction** 1

1.1   Quarks 1
1.2   Colour 2
1.3   Flavour 2
1.4   Quark Masses 3
1.5   Medium Effects on Quarks and Their Interactions 4
1.6   Phase Transition and Quark Gluon Plasma 6

2   **Preliminaries and Simple Models** 9

2.1 $SU(2)$ and $SU(3)$ Symmetry 9
2.2   The $SU(3)$ Group 10
2.3   Non-Abelian Gauge Transformations 11
2.4 $SU(3)$ Flavour Group 12
2.5   The $SU(3)$ Colour and the Multi-Quark Wave Function 13
2.6   The Necessity of Relativistic Models 15
2.7   Fully Relativistic One-Body Potential Models 19
2.8   Relativistic Hartree-Fock Models 20

3   **Currents, Anomaly, Solitons and Fractional Fermions** 24

3.1   The Theorem of Emmy Noether 24
3.2   Internal Symmetry and Space-Time Symmetry 25
3.3   Standard Examples 25
3.4   Current Algebra and Anomaly 26
3.5   Fractional Fermions 28
3.6   A Brief Review of Solitons: Kinks 29
3.7   Polyacetylene: One Dimensional Dirac-Type Equation 30

4   **More on Chiral Anomaly** 32

4.1   Chiral Anomaly in QED 32
4.2   Gentleness 32

4.3   Momentum Routing Anomalies 34
4.4   Anomalous Ward Identity from Path Integral Approach 36
4.5   Summary of Grassmann Algebra 37
4.6   Integration Over Fermi Fields 39
4.7   Chiral Transformation 39

5   Introduction to Instantons 43

5.1   Instantons or Pseudoparticles 43
5.2   Tunneling in Imaginary Time 44
5.3   Homotopy Theory 46
5.4   Compactification of Space-Time Manifold and Mapping 47
5.5   Bounds for the Instanton Solutions 50

6   Relevance of Instantons 53

6.1 θ Vacuum 53
6.2   The DEMON 54
6.3   Electroweak Baryon Number Violation 56
6.4   Conformally Invariant Solutions of Jackiw, Nohl and Rebbi 56
6.5   Instanton to Skyrmion 59
6.6   QCD Vacuum – Instanton Gas or Liquid? 63
6.7   Instanton Suppression by Light Fermions 64
6.8   Instanton Induced Effective Interaction 65

7   Chiral Perturbation Theory (CHPT) 72

7.1   CHPT for the Meson Sector 72
7.2   CHPT for the Nucleon 76
7.3   CHPT at Finite Temperature 77

8   The Topological and Non-Topological Soliton Models 79

8.1   Skyrmion: Baryon as a Topological Soliton in Meson Fields 79
8.2   The Non-topological Soliton or the Soliton Bag Model 81
8.3   The Nature of the QCD Vacuum 81
8.4   Description of the Model 81

9   QCD Sum Rules 83

9.1   Introduction to QCD Sum Rules 83
9.2   The Operator Product Expansion (OPE) 85

9.3 Calculation of the Coefficients and Borel Transform 87
9.4 Introduction to Sum Rules for Baryons 92
9.5 The Nucleon Correlator 92
9.6 Details of the Sum Rule for Nucleons 93
9.7 Finite Temperature and Density 94

References 97

Subject Index 105

Citation Index 107

1 Introduction

1.1 Quarks

Quarks (q) and antiquarks ($\bar{q}$) are confined particles in hadrons. The hadrons (baryons, N, Λ, Σ, Δ ..., and mesons, π, ρ, ω, K ...) are directly observed, whereas the q and the $\bar{q}$ are not. There may be another phase, and we shall talk about this later, in which q and $\bar{q}$ are deconfined but we do not live in that phase. So our instruments cannot experimentally detect them in that phase. When they recombine into observable hadrons they are again in the confining phase. So the existence of this deconfined phase must be a pure theory (and it better be a good reliable theory) or they must leave some mark, some *signature* of their passage into this deconfined phase. Are there such signatures? Experimentalists and theorists all over the world are eagerly searching for such signatures. It is an exciting branch of physics for experiments and also for theory (supercomputers, nationwide computer link-ups etc.). The suppression of the production of an exotic meson J/ψ, (also called gypsy!) in heavy ion collisions may be a candidate. This was predicted from theory by *Matsui* and *Satz* (1986) and found in experiments. But there are other possible explanations of this from nuclear physics. This gypsy suppression is typical and we shall discuss it in this chapter, but not with too much detail.

The presence of q and $\bar{q}$ can also be inferred indirectly from deep inelastic scattering experiments (DIS). This is inelastic lepton (electron, muon, or neutrino) scattering from hadrons, with a very large momentum transfer q^2. The lepton interacts through exchange of the photon γ or W, Z mesons according to known laws of electroweak theory. It is found from the experiments that much of the missing momentum of the inelastic process is carried away by *pointlike* particles which have no structure of its own. These are called partons (*Close* 1979). There appear to be two types of partons: (i) a large part of the momentum is carried away by gluons, spin 1, chargeless, massless particles. (ii) The other part of q^2 is shared by spin $1/2$ fermions, believed to be quarks and antiquarks, some real and some virtual. It so turns out that for the theory of q, $\bar{q}$ and gluons, called quantum chromodynamics QCD, one can do perturbation theory at high q^2 where the interactions are small. This makes analyses of DIS experiments very interesting, but here we will not be concerned with the detailed theory and refer the reader to *Close* (1979) and *Muta* (1987). But at low q^2, which corresponds to interactions which are long-ranged, quarks (antiquarks) interact via gluons and the strong long-ranged interaction cannot be summed in a perturbation series. Part of the problem is due to the fact that although QCD is modelled on QED, where gauge invariance leads to nice renormalizable perturbation results, there is an extra degree of freedom called colour.

1.2 Colour

Quarks (antiquarks) have colour and observable hadrons are colourless, or colour singlet. Colour is the strong interaction charge which comes in three forms. Unlike QED where photons are charge neutral, in QCD gluons carry colour indices and can interact with themselves. A q (or $\bar{q}$) will change colour when it emits a gluon.

The colour degree of freedom belongs to a group much like spin. For spin, the Pauli matrices are the generators of the $SU(2)$ group. The colour group is larger – $SU(3)$, with eight generators, but often lattice theorists work with a hypothetical $SU(2)$ colour group for simplicity. We refer the reader to Section 2.1 for more about the $SU(3)$ group. A more interesting generalization of this group structure is to take it to be $SU(N)$ where N is large. Of course 3 is not a large number but if one expands the theory in powers of $1/N$, the leading term after unity is about 30%. Results good upto 30% are very encouraging for physics of strong interactions where one cannot do perturbation theory. Of course there is the special case of DIS, where perturbation theory is permitted, as already mentioned. But for most other cases we need to go out of the strait-jacket of perturbation theory. We will be mostly concerned with these techniques: mean field theory, QCD sum rules using operator product expansion (OPE in short) or solitons and instantons.

1.3 Flavour

Apart from colour, quarks have another quantum number called flavour. Flavour comes in families of pairs matching quarks with leptons. This is required by renormalizability and unitarity of the Glashow-Weinberg-Salam theory of electroweak processes through a phenomenon of anomaly cancellation which we shall discuss further later on. The first family (u, d) is paired with the electron e and the corresponding neutrino (ν_e) and is the $SU(2)$ subgroup of the flavour $SU(N_F)$, relevant for low energy. This is the $SU(2)$ isospin subgroup with which the nuclear physicists are familiar. In physical terms, the near equality of the masses of the proton and neutron or of the three pions π^+, π^- and π^0 is due to the near equality of the masses of the (u, d) quarks and this is isospin symmetry. In each family of quarks and leptons, the sum of the charges must be zero, – so if the quarks have fractional charges they must occur in three colours. For example, charges for u and d are $(2/3)$ and $(-1/3)$ of the electronic charge. To cancel with the charge of (-1) we need 3 colours. For details of this anomaly cancellation we refer the reader to *Pokorski* (1987, p. 613).

Next comes the strange and the charm quark. The latter is the constituent of the J/Ψ particle which was predicted through the GIM mechanism (*Glashow, Illiopoulos* and *Maiani*, 1974; this is something to do with strangeness changing currents, we refer the reader to *Itzykson* and *Zuber*, 1980, p. 628). These quarks are associated with the μ-meson and its neutrino (ν_μ).

The beauty or bottom quark was discovered in $Y(\bar{b}b)$. The partner of the *b*-quark, the so-called top, have not been found yet. The lepton pair is the heavy electron τ and ν_τ. Charges are $(2/3)$ for *c* and top quark and $(-1/3)$ for *s* and *b* quark. The proton consists of valence quarks *uud*, neutron-*udd* and the hyperon Λ-*uds* etc.

1.4 Quark Masses

So far we have discussed the quantum numbers of the quarks which are straight forward. In the past one had thought of other types of quarks: the Sakata triplet or the Han-Nambu nonet. Now we believe in coloured quarks of various flavours.

There is evidence from DIS of electrons on hadrons that for large q^2, *u* and *d* quarks are almost massless (4 and 7 MeV) and the *s* quark has a mass of about 150 MeV (*Gasser* and *Leutwyler*, 1982). These are the current masses and if we consider these quarks only we can go to the chiral limit of QCD where the quark masses are neglected. At large q^2 one also has very little interaction between quarks and so a massless weakly interacting quark gas is a good description of a system. But at small q^2 the quarks interact very strongly and the chiral limit is no longer valid. We say the chiral symmetry is broken. Usually, associated with such breaking is the existence of a massless boson, the Nambu-Goldstone boson. In this case it is the pion. In the real world of course the u, d quarks also have a small mass and the pion mass is proportional to this. It is possible to consider the quark mass to be dependent on the distance of the probe *r* from its centre and such pictures of co-ordinate dependent mass $m(r)$ will be discussed in Chapter 2. It may be that by the time a probe is as far away as 0.1 fm from the quark, it sees a dressed u, d mass of about 300 MeV. These are the so-called constituent masses used in non-relativistic oscillator models.

We will define chiral symmetry and the left and right handed quarks at this stage. We start with the Lagrangian density given in terms of the four-component spinors ψ and the mass of the fermion

$$\mathcal{L} = i\bar{\psi}\partial_\mu\gamma_\mu\psi - m\bar{\psi}\psi \tag{1.4.1}$$

The Dirac matrices γ_μ are given below (*Bjorken* and *Drell*, 1964):

$$\gamma^i = \begin{pmatrix} 0 & \alpha_i \\ -\alpha_i & 0 \end{pmatrix}, \quad i = 1, 2, 3 \quad \gamma^0 = \begin{pmatrix} I & 0 \\ 0 & -I \end{pmatrix}, \tag{1.4.2}$$

$$\gamma_5 = i\gamma_0\gamma_1\gamma_2\gamma_3 = \begin{pmatrix} 0 & I \\ I & 0 \end{pmatrix} = \gamma^5 \tag{1.4.3}$$

We define (left, right) handed spinors in terms of γ_5 as follows

$$\psi_L = 1/2(1 - \gamma_5)\psi,$$
$$\psi_R = 1/2(1 + \gamma_5)\psi,$$
$$\psi = \psi_L + \psi_R \tag{1.4.4}$$

It is easy to check that

$$\bar{\psi}\psi = \bar{\psi}_L\psi_R + \bar{\psi}_R\psi_L \tag{1.4.5}$$

so that we see that the mass term in Eq. (1) mixes the left and the right fermions whereas one can also check easily that

$$\bar{\psi}\partial_\mu\gamma_\mu\psi = \bar{\psi}_L\partial_\mu\gamma_\mu\psi_L + \bar{\psi}_R\partial_\mu\gamma_\mu\psi_R. \tag{1.4.6}$$

If we now think of a global rotation, introducing a phase in ψ as follows

$$\psi' = \exp(i\gamma_5\theta)\psi \tag{1.4.7}$$

called a chiral rotation or axial rotation then the Lagrangian density given in Eq. (1) remains invariant provided the mass m is zero. Since for the (u, d) current quarks the masses are often negligible there is a chiral symmetry for (u, d) quarks. One can start neglecting them and then build up a perturbation theory introducing the small masses. This is chiral perturbation theory which is very powerful. This was first introduced by Weinberg and has been developed by Leutwyler and his group (Chapter 7).

1.5 Medium Effects on Quarks and Their Interactions

We now discuss why nuclear physics becomes important for getting at the properties of these elusive quarks. We have already indicated that although the QCD Lagrangian is roughly chirally invariant at low energy, this invariance is broken in the ground state and the pion is generated as the Goldstone boson of this spontaneous breaking of chiral symmetry (χSB or SBCS in short, see Chapter 5–6). Low energy nuclear physics is dominated by SBCS. This is because, essentially, the gluons do not mix in too much of the heavier quarks with u, d (*Applequist* and *Carrazone*, 1975). But the strange quark is, of course, light enough to be mixed in. So recently nuclear physicists have realized that there may be non-trivial consequences of this mixing (see for example *Brown* et al. 1988).

If the ambient quark density of a system increases, as for example in a heavy ion (HI) collision, it may be that the chiral symmetry is restored. One looks for such events in nuclear collisions. Furthermore, the confining potential between quarks, for example the c and the $\bar{c}$ in case of the J/Ψ, may decrease because of Debye screening, a term used in solid state theory. In this case these particles are not held together and one should observe less J/Ψ in HI collisions. But as pointed out by numerous authors, the Matsui-Satz effect may be masked by inelastic collisions of the J/Ψ with nucleons in the medium, whereby other particles are formed.

In general HI collisions are complicated by final state interactions and it is hard to single out one particular medium-dependent effect from another. In a gas of hadrons, J/Ψ will interact with other mesons and nucleons. This has been the subject of many theoretical papers. For theoretical details we refer to *Satz* (1988). *Nagamiya* (1988) has reviewed the experimental situation. He quotes the ratio of J/Ψ to continuum as:

$$N_\psi/N_c = 9.3 \pm 0.6, \text{ for peripheral}$$

$$= 5.9 \pm 0.4, \text{ for central collisions} \tag{1.5.1}$$

from the NA38 group at CERN, implying a reduction in central collisions as compared with peripheral. According to Nagamiya the 125 GeV pion + A collisions show that on changing the Beryllium target to Tungsten the J/Ψ yield is decreased by 20–50%. The experimental results are from the FNAL group. This may be due to the extra nucleons in Tungsten compared to light Be, and as he comments: most probably this suppression is due to final state interactions.

We will next talk about the EMC effect, named after the European Muon Collaboration working at CERN, who found from DIS on iron target by high energy muon projectiles that the distribution of quarks in iron is not the same as in deuteron. *Close, Roberts* and *Ross* (1986) had tried to explain this as an increase in confinement size and other groups had argued that this is due to enhanced pion or nuclear binding effects. The problem is reviewed beautifully in Close, Roberts and Ross where they discuss the duality between their "swelling of the nucleon" and the idea that the "degradation of the valence quarks" is due to energy-momentum transfer to the meson degrees of freedom. The scaling of nucleon and meson properties with density is an interesting area of study (*Dey* and *Dey*, 1986; *Dey* et al. 1988; *Brown* and *Rho*, 1990, 1991).

The EMC group came up with another piece of startling data on polarized leptons off polarized nucleons in deep inelastic scattering which has received an enormous amount of attention of late. From the experiment one extracts the lowest moment of the polarized structure function for the proton.

A first interpretation of these data seems to suggest that the total amount of spin carried by the quarks in a proton is essentially zero. In terms of quark distribution $q_+(q_-)$ as a function of Bjorken variable x, the quark densities Δq are:

$$\Delta q = \int_0^1 dx(q_+ + \bar{q}_+ - q_- - \bar{q}_-) \tag{1.5.2}$$

They are related to matrix elements of the proton p of spin s, associated with the axial vector currents

$$\Delta q S_\mu = \langle p, s|\bar{q}\gamma_\mu\gamma_5 q|p, s\rangle \tag{1.5.3}$$

where S_μ is the proton spin axial four-vector. Using $SU(3)$ and fixing the coupling coefficients F and D defined by the axial vector matrix elements of the members of the baryon octets from global fits to hyperon decays (*Fritzsch*, 1990):

$$F = 0.47 \pm 0.04, \quad D = 0.81 \pm 0.03$$

$$\Delta u = 0.74 \pm 0.10$$

$$\Delta d = -0.54 \pm 0.10$$

$$\Delta s = -0.20 \pm 0.11 \qquad\qquad (1.5.4)$$

The sum is nominally equal to zero!

The evidence seems to suggest that mesons, strange sea quarks or gluons (or perhaps both) are very important in the proton and carry much of the spin, either as spin or as orbital angular momentum. We will also talk about this in Chapter 6 in connection with instantons, which are some gluon configurations.

Other approaches are also possible. A recent example is the suggestion by *Jaffe* and *Lipkin* (1991) that the nucleon may not be a purely three quark state $3q$, but is partly a $3q$ and meson combination! The allowed colour singlet meson combinations are $f_0(0^{++})$ and $f_1(1^{++})$. The model explains hyperon beta decays and baryon magnetic moments and shows that the nucleon wave function can be extended simply to accommodate the EMC result. We pass over now to other topics.

Why are nucleus-nucleus collisions better suited to producing a quark-gluon plasma than ordinary hadron collisions? The obvious reason is that in nucleus-nucleus collisions, the average multiplicity of produced particles is higher than for hadron-hadron collisions, and higher energy densities result on the average. Another is that nuclei are of large spatial extent, and in order to produce a well-thermalized matter distribution, the particles produced in the collisions must rescatter, so that the large spatial size enables the produced particles to rescatter several times before being emitted from the collision region. For description of collisions and also phase transitions which we will discuss in the next section, we need thermodynamical concepts.

Why is statistical mechanics used to study strongly interacting systems? The fact that strong interactions (when the interactions are truly strong as opposed to the perturbative large q^2 domain) are intimately connected to multiparticle dynamics seems to be an inevitable consequence of the particle number non-conserving nature of such interactions. Even the vacuum of QCD is a state consisting of an infinite number of virtual quarks, antiquarks and gluons. Statistical mechanical techniques seem ideally suited for analyzing such systems.

1.6 Phase Transition and Quark Gluon Plasma

The question of phase transition is a very tricky one. There is no direct experimental confirmation of a possible phase transition either at high temperature or at high density. As we have indicated the experimental signatures are unclear because of complications like final state interaction etc. For the pure gauge sector (i.e. for gluons only), lattice calculations gave a first order phase transition, thus defining a temperature T_c for such calculations. The addition of quarks in lattice theories makes the calculation much less reliable. It was believed, however, that there is a first order chiral phase transition, when the approximate chiral symmetry of QCD is restored. It was also hoped that the deconfinement phase transition, where the quarks and gluons become like a free quark-gluon plasma (QGP in short), also exists and that the two transitions may take place together. From chiral perturbation theory (CHPT), for example, one could extract the temperature to be around 200 MeV. But as we shall see in Chapter 7, CHPT is a low temperature expansion and its prediction at $T = 200$ MeV need not be taken too literally. *De Tar* and *Kogut* (1987) found, in lattice calculations, that there is no deconfinement but rather the occurrence of parity doublets and *large hadrons* at high temperatures. Measurements carried out at temperatures just above T_c give evidence of large residual interaction. For example, the pressure and energy densities have been looked at by *Engels* et al. (1990). One can calculate the quark number susceptibility and show that quarks are physically present participating in a non-local four fermion interaction (*Gottleib* et al. 1987b, *Gavai, Potvin* and *Sanielevici* 1989).

The parity doublets found by De Tar and Kogut have been confirmed by *Gottlieb* et al. (1987a, 1990) and by many others (*Gocksch, Rossi* and *Heller* 1988; *Ukawa* 1989; *Fukugita* et al. 1989, 1990; *Gavai* et al. 1990; *Born* et al. 1991; *Brown* et al. 1991; *Kogut* et al. 1991). The question of large hadrons is more subtle since all lattice calculations have finite size to keep the computation manageable, and *Gavai* et al. (1990) for example find the correlation length for the large hadrons comparable to the limit on their correlation length. But it appears there *is no first order deconfinement phase transition* and there is no large latent heat associated with such transitions with two generations of almost massless quarks. Let us digress a bit on these interesting and very recent topics.

In finite temperature field theory there are two types of masses: one which appears as the pole of the propagator (the Green's function) of the field, and the other appears in the Yukawa type fall off of the correlators of the currents. The first is sometimes called the pole mass and the second is often known as the screening mass. Of course, both go over to the unique physical mass, in the zero temperature limit.

Lattice calculations are done in Euclidean space-time and temperature (T) introduces periodicity in the time axis as we shall discuss in Chapter 6 Section 5b. The masses measured by the lattice groups are the screening masses at finite temperature. It is found that if there was no interaction for the quarks (presumably this is what would happen at very high T) then the self energy of quarks would be screened like $\exp(-\pi T)$. Then, in this scenario, the meson screening masses would go as $2\pi T$ and the baryons as $3\pi T$. Indeed this is what is found in the lattice calculations for vector/axial vector mesons and baryons. Obviously since there is no interaction, the odd and even parity states appear together. In case of the scalar/pseudoscalar channels however the screening masses are less than $2\pi T$, although parity doubling is still seen. The states are still confined at T at which the pure glue phase transition takes place. The confinement is only in the transverse correlator and apparently (*Koch* et al. 1992) even a weak force can produce such confinement. Modelling these results will be a major concern for physicists in the near future. For example there is an interesting attempt by *Gupta* (1992) who finds that for the pseudoscalar channel one can find a four fermion interaction that fits the lattice data, whereas the interaction is almost zero in the ρ-meson channel.

2 Preliminaries and Simple Models

2.1 $SU(2)$ and $SU(3)$ Symmetry

Nuclear physicists are familiar with isospin symmetry, the $SU(2)$ symmetry between the neutron and the proton. This was generalized to include strange particles, introducing strangeness or hypercharge to extend the symmetry to flavour $SU(3)$. In quark language this implies assuming a symmetry between the u, d and the strange quark s. This symmetry is not exact, but is broken since the s-quark is heavier.

(u, d) form the fundamental representation for $SU(2)$ flavour group. The generators $T_i \equiv \frac{1}{2}\tau_i$ satisfy $[T_i, T_j] = i\varepsilon_{ijk}T_k$ and the τ_i's are the mathematical copies of the Pauli matrices:

$$\tau_1 = \begin{pmatrix} 0 & 1 \\ 1 & 0 \end{pmatrix} \qquad \tau_2 = \begin{pmatrix} 0 & -i \\ i & 0 \end{pmatrix} \qquad \tau_3 = \begin{pmatrix} 1 & 0 \\ 0 & -1 \end{pmatrix} \tag{2.1.1}$$

The anti-particle doublet can be obtained as follows. Consider a π rotation in isospin space around the 2nd axis:

$$\begin{pmatrix} u' \\ d' \end{pmatrix} = \exp(-i\pi T_2) \begin{pmatrix} u \\ d \end{pmatrix} = -i\tau_2 \begin{pmatrix} u \\ d \end{pmatrix} = \begin{pmatrix} 0 & -1 \\ 1 & 0 \end{pmatrix} \begin{pmatrix} u \\ d \end{pmatrix} \tag{2.1.2}$$

Anti-particle states are obtained from the charge conjugation operator: $Cu = \bar{u}$ and $Cd = \bar{d}$. But operation by C on (2) is not enough. Generally, the most positively charged particle is chosen to have the maximum T_3. Hence the reordering of $\bar{u}$ and $\bar{d}$ with a negative sign in $\bar{d}$ are required so that the doublet:

$$\begin{pmatrix} -\bar{d}' \\ \bar{u}' \end{pmatrix} = \begin{pmatrix} 0 & -1 \\ 1 & 0 \end{pmatrix} \begin{pmatrix} -\bar{d} \\ \bar{u} \end{pmatrix} \tag{2.1.3}$$

transforms identically like (2) : $\bar{d}' \to \bar{u}, \bar{u}' \to -\bar{d}$ like the particles $u' \to -d$, $d' \to u$. This is a special property of $SU(2)$.

A composite system of quark-antiquark pair (mesons) has the isospin states in irreducible representations 3 and 1 originating from $2 \otimes 2 = 3 \oplus 1$:

$$
\boxed{
\begin{aligned}
|T = 1, \quad T_3 = 1\rangle &= -u\bar{d} \\
|T = 1, \quad T_3 = 0\rangle &= (u\bar{u} - d\bar{d})/\gamma_2 \\
|T = 1, \quad T_3 = -1\rangle &= d\bar{u}
\end{aligned}
}
\qquad
|T = 0, \quad T_3 = 0\rangle = (u\bar{u} + d\bar{d})/\gamma_2
\tag{2.1.4}
$$

2.2 The $SU(3)$ Group

The set of Unitary 3×3 matrices with Det $U = 1$ form the $SU(3)$ group. There are $3^2 - 1 = 8$ generators and they may be taken as any 8 linearly independent traceless 3×3 hermitian matrices. Since only 2 out of 8 such matrices may be made diagonal, this is the maximum number of mutually commuting generators. This number is called the rank of the group and it is the same as the number of Casimir operators.

A standard representation of the generators $T^\alpha \equiv \lambda^\alpha/2$ are given by the eight λ-s known as the Gell-Mann matrices:

$$\lambda_1 = \begin{pmatrix} 0 & 1 & 0 \\ 1 & 0 & 0 \\ 0 & 0 & 0 \end{pmatrix}, \qquad \lambda_2 = \begin{pmatrix} 0 & -i & 0 \\ i & 0 & 0 \\ 0 & 0 & 0 \end{pmatrix},$$

$$\lambda_3 = \begin{pmatrix} 1 & 0 & 0 \\ 0 & -1 & 0 \\ 0 & 0 & 0 \end{pmatrix}, \qquad \lambda_4 = \begin{pmatrix} 0 & 0 & 1 \\ 0 & 0 & 0 \\ 1 & 0 & 0 \end{pmatrix},$$

$$\lambda_5 = \begin{pmatrix} 0 & 0 & -i \\ 0 & 0 & 0 \\ i & 0 & 0 \end{pmatrix}, \qquad \lambda_6 = \begin{pmatrix} 0 & 0 & 0 \\ 0 & 0 & 1 \\ 0 & 1 & 0 \end{pmatrix},$$

$$\lambda_7 = \begin{pmatrix} 0 & 0 & 0 \\ 0 & 0 & -i \\ 0 & i & 0 \end{pmatrix}, \qquad \lambda_8 = \frac{1}{\sqrt{3}} = \begin{pmatrix} 1 & 0 & 0 \\ 0 & 1 & 0 \\ 0 & 0 & -2 \end{pmatrix}, \qquad (2.2.1)$$

These λ-commutators satisfy the group equation

$$[\lambda_k/2, \lambda_l/2] = if_{kln}\lambda_n/2 \qquad (2.2.2a)$$

where the structure constants f are totally antisymmetric in their indices and their values are:

$$f_{123} = 1, \ f_{147} = -f_{156} = f_{246} = f_{257} = f_{245} = f_{267} = \frac{1}{2}$$

$$f_{458} = f_{678} = \frac{\sqrt{3}}{2} \qquad (2.2.2b)$$

Also

$$if^{abc} = 1/4 Tr(\lambda^a\lambda^b\lambda^c - \lambda^b\lambda^a\lambda^c), \ f^{abc} \equiv f_{abc} \qquad (2.2.2c)$$

In $SU(N)$ the dimension of the fundamental representation is N and that of the adjoint representation is $N^2 - 1$:

$$Tr[T^a, T^b] = \frac{1}{2}\delta^{ab}$$

$$[T^a_{adj}, T^b_{adj}] = if^{abc}T^c_{adj}$$

$$(T^a_{adj})_{bc} = -if_{abc} \text{ in } SU(3) \qquad (2.2.3)$$

$(T^a_{adj})_{bc} = -i\varepsilon_{abc}$ in $SU(2)$. In particular, the pion triplet in $SU(2)$ and eight coloured gluons in $SU(3)$ transform like the adjoint. From the first three λ-s, it is clear that $SU(2)$ is a subgroup of $SU(3)$. The fundamental representation of $SU(3)$ is a triplet. In $SU(3)$ colour it is a triplet of (r, g, b) and for $SU(3)$ flavour it is the (u, d, s).

2.3 Non-Abelian Gauge Transformations

If there exists a symmetry in a theory the Lagrangian must remain invariant under the group transformations. Consider the matter field in QED:

$$\mathcal{L} = \bar{\psi}(x)(i\gamma_\mu \partial_\mu - m)\psi(x) \tag{2.3.1}$$

The only parameter of $U(1)$ gauge transformation is, say α. $\mathcal{L}$ is invariant under $U(1)$: $\psi' \rightarrow e^{i e \alpha} \psi$; $\bar{\psi}' \rightarrow \bar{\psi} e^{-i e \alpha}$. If this symmetry is a local one $\alpha \rightarrow \alpha(x)$ we need gauge fields $A_\mu(x)$. For $U = e^{i e \alpha(x)}$, we have

$$\psi'(x) \rightarrow U\psi(x), \bar{\psi}'(x) \rightarrow \bar{\psi}U^{-1}, A'_\mu \rightarrow UA_\mu U^{-1} - \frac{i}{e}(\partial_\mu U)U^{-1} \tag{2.3.2}$$

$\partial_\mu \psi'$ picks up an extra term $i e \partial_\mu \alpha(x)$ which is cancelled from A'_μ. The invariant Lagrangian becomes:

$$\mathcal{L} = \bar{\psi}(x)(i\gamma_\mu D_\mu - m)\psi(x) - \frac{1}{4}F_{\mu\nu}F^{\mu\nu} \tag{2.3.3}$$

where the covariant derivative $D_\mu = \partial_\mu - ieA_\mu$ and the field tensor $F_{\mu\nu} = \partial_\mu A_\nu - \partial_\mu A_\nu$. They transform as

$$D_\mu \psi \rightarrow (D_\mu \psi)' = U(D_\mu \psi); F'_{\mu\nu} = UF_{\mu\nu}U^{-1} \tag{2.3.4}$$

For non-Abelian $SU(3)$ symmetry the group element is an 8-parameter 3×3 matrix, transforming the fields as $\varphi' \Rightarrow U(x)\varphi \equiv e^{-igT_i \cdot \theta_i(x)}\varphi$ where φ is a triplet in internal space. The covariant derivative and matrix-valued vector fields are:

$$D_\mu = \partial_\mu - igA_\mu; A_\mu(x) = A^a_\mu T^a; F_{\mu\nu} = F^a_{\mu\nu}T^a \tag{2.3.5}$$

A_μ transforms as (2) and D_μ and $F_{\mu\nu}$ transform as Eq. (4). For infinitesimal transformation

$$U_{kl} = \delta_{kl} - igT^a_{kl}\theta^a; U^{-1}_{kl} = \delta_{kl} + igT^a_{kl}\theta^a \tag{2.3.6}$$

the Eq. (2) becomes: $T^a_{kl}A^{a'}_\mu = T^a_{kl}A^a_\mu - ig\theta^b[T^b, T^a]_{kl}A^a_\mu - T^a_{kl}\partial_\mu\theta^a$, hence

$$\delta A^a_\mu = gf_{abc}\theta^b A^c_\mu - \partial_\mu\theta^a \tag{2.3.7}$$

The gauge field tensor (5) satisfies $[D_\mu, D_\nu]\varphi = -igF_{\mu\nu}\varphi$ with

$$F^a_{\mu\nu} = \partial_\mu A^a_\nu - \partial_\nu A^a_\mu + gf_{abc}A^b_\mu A^c_\nu \tag{2.3.8}$$

Under infinitesimal transformations (6), Eq. (4) gives

$$\delta F^a_{\mu\nu} = gf_{abc}\theta^b F^c_{\mu\nu} \tag{2.3.9}$$

which means $F_{\mu\nu}^c \Rightarrow (U_{adj})^{cd} F_{\mu\nu}^d$ (see the Eq. (2.4) for the adjoint). So unlike the photons in QED, the gluons in QCD couple with themselves and as a result $F_{\mu\nu}$ is *gauge covariant, not invariant* and can be contracted with itself to give the invariant $\mathcal{L}$:

$$\mathcal{L} = \bar{\psi}(x)(i\gamma_\mu D_\mu - m)\psi(x) - \frac{1}{2} Tr F_{\mu\nu} F^{\mu\nu} \qquad (2.3.10)$$

2.4 $SU(3)$ Flavour Group

We shall not go into details of the exhaustively studied $SU(3)$ flavour symmetry. The convention is to define a quantum number $Y \equiv$ Baryon number + Strangeness $= B+S$ and then the charge $Q = T_3 + Y/2$. The quantum numbers are tabled in 2.1. With B-number conservation one may create or destroy quark-antiquark $(q\bar{q})$ pairs but not a single quark, q. The former process may be suppressed though, since a soft gluon exchange is involved. In strong and electro-magnetic interactions q keeps its identity. Transmutations like $s \Rightarrow u+$ leptons is possible in weak interaction. The $SU(2)$ states (Eq. (2.1.4)) are identified for example with 3 pions and η meson. They are embedded in to the $SU(3)$ group of u, d and s. The generalization is $3 \otimes \bar{3} = 1 \oplus 8$. The extra states in *octet* are the $K^\pm$, K^0 and $\bar{K}^0$. The $s\bar{s}$ will not contribute to $T = 1$ but will mix with $(u\bar{u} + d\bar{d})/\sqrt{2}$, the $T = 0$ state, to give the $SU(3)$ singlet η' and the octet η. Masses in the same multiplet should be the same if there is an exact $SU(3)$ symmetry which is not the case because the u, d masses are different from the s-mass. This may explain the $\pi - K$ mass difference but $\eta - \eta'$ mass difference is more serious and involves instantons, which we will discuss later in Chapters 5 and 6 and also in the last chapter.

We write down 27 combinations of 3 qqq triplet. They arrange themselves into 4 irreducible representations (see for example *Close*, 1979):

$$3 \otimes 3 \otimes 3 = 10 \oplus 8_1 \oplus 8_2 \oplus 1 \qquad (2.4.1)$$

They are the symmetric decuplet, mixed symmetric 8_1 and mixed antisymmetric 8_2 octets and the antisymmetric singlet. For describing the spin $1/2$ angular momentum, $SU(2)$, u and d are replaced by spin up and spin down projections, α and β, respectively. 3 quark doublets will give rise to the following arrangements:

$$2 \otimes 2 \otimes 2 = (3 \oplus 1) \oplus 2 = 4 \oplus 2 \oplus 2 \qquad (2.4.2)$$

Table 2.1. Quantum numbers of the quarks

Quark	Spin	B	Q	I_3	S	Y
u	$1/2$	$1/3$	$2/3$	$1/2$	0	$1/3$
d	$1/2$	$1/3$	$-1/3$	$-1/2$	0	$1/3$
s	$1/2$	$1/3$	$-1/3$	0	-1	$-2/3$

3 and 1 in the brackets are the symmetric and antisymmetric spin – 1 and 0 respectively (analogous to Eq. (1.4)). Clearly, 4 corresponds to the most symmetric spin $3/2$, 2 the mixed symmetric one with spin $1/2$ and the last 2, mixed antisymmetric with spin $1/2$. We denote them by S, λ and ρ. For spin $3/2$ with projection $1/2$, the S wavefunction is given by:

$$\chi^s = \frac{1}{\sqrt{3}}(\alpha\,\alpha\,\beta + \alpha\,\beta\,\alpha + \beta\,\alpha\,\alpha) \tag{2.4.3}$$

where the first term implies $\alpha(1)\,\alpha(2)\,\beta(3)$ and so on. In the same notation one can have the two other possibilities

$$\chi^\lambda = \frac{1}{\sqrt{6}}(2\alpha\,\alpha\,\beta - \alpha\,\beta\,\alpha - \beta\,\alpha\,\alpha) \tag{2.4.4}$$

$$\chi^\rho = \frac{1}{\sqrt{2}}(\alpha\,\beta - \beta\alpha)\alpha \tag{2.4.5}$$

The isospin wave functions denoted by φ^s, φ^λ, φ^ρ are exactly the same. But 3-flavour wave functions are more complicated. For example, we get the antisymmetric states also, denoted by A as shown in Eq. (1).

To construct baryon in a quark model, combine $SU(3)$ flavour and $SU(2)$ spin (Eqs. (1) and (2)) according to permutation symmetry:

$(10 \oplus 8 \oplus 8 \oplus 1), (4 \oplus 2 \oplus 2) \rightarrow$

$S : (10, 4) + (8, 2) \rightarrow 56 - \text{plet}$

$\lambda : (10, 2) + (8, 4) + (8, 2) + (1, 2) \rightarrow 70 - \text{plet}$

$\rho : (10, 2) + (8, 4) + (8, 2) + (1, 2) \rightarrow 70 - \text{plet}$

$A : (1, 4) + (8, 2) \rightarrow 20 - \text{plet}$ (2.4.6)

2.5   The *SU(3)* Colour and the Multi-Quark Wave Function

There is a major difference between nuclear (and atomic) physics wave functions and the quark wave function in a baryon. The antisymmetry of the wave function is ensured by the colour part and the spin-flavour-space part is symmetric. The colour wave function can be represented simply as follows

$$\frac{1}{\sqrt{6}}\varepsilon_{abc}\xi_a\xi_b\xi_c \tag{2.5.1}$$

a, b, c correspond to colour indices r, g and b and sum is implied over repeated indices. The gluon exchange operator in colour space is a singlet that can be represented by

$$\lambda_1^{pq}\lambda_2^{rs} \tag{2.5.2}$$

where the indices (p and q for example) imply the destruction of a particular colour and creation of another. Since this is a two-body interaction the colour of the third particle remains intact and the matrix element of the interaction Eq. (2) between two states of type given in Eq. (1) can be written as

$$\frac{1}{6}\varepsilon_{pro}\lambda_1^{pq}\cdot\lambda_2^{rs}\varepsilon_{qso} = \frac{1}{6}(\delta_{pq}\delta_{rs} - \delta_{ps}\delta_{qr})\lambda_1^{pq}\cdot\lambda_2^{rs} \qquad (2.5.3)$$

when the sum over the index c is performed. The first term gives $(tr\lambda)^2$, which is zero, implying it is not possible to keep the colour of the particle the same while exchanging a gluon. The second term gives $tr\,(\lambda^2)$ and is negative implying attraction. This example illustrates how colour factors are to be considered. Also it shows that from one gluon exchange (OGE) one expects an attractive force. The final result is $8/9$. For $q\bar{q}$ mesonic states the value is twice as much. In early quark models the confining force between 2-quarks in a baryon was also taken to be half of that between a quark and an antiquark in a meson. But the confining force is not generated by a simple OGE. So later this factor became an object of study (see *Lucha* et al. 1990). *Capstick* and *Isgur* (1986) took it to be $\cong 0.55$ from the flux tube model.

The space wave functions for 3 quarks can also be arranged according to permutation symmetry into symmetric, mixed and antisymmetric states. With oscillator wavefunctions *Isgur* and *Karl* (1978, 1979) worked out the spectroscopy in details. From Eq. (4.6) it is clear that the ground state baryons ($L = 0$) will form a 56-plet. The lowest mass N, Λ, Σ and Ξ fit neatly into octet and Δ, Σ^*, Ξ^* and Ω into the decuplet. The first excited states are a bit more complicated and one can show that ρ- and λ-type excitations ($L = 1$ wavefunctions) combine with appropriate spin-flavour to form a 70-plet. The $L = 2$ states have both 56-plet and 70-plet along with a space-antisymmetric 20-plet.

The analogue of the Fermi-Breit interaction for electrons is present for quarks in the non-relativistic limit for one gluon exchange force. This gives the octet-decuplet splitting as was shown by *de Rujula*, *Georgi* and *Glashow* (1975). The $N - \Delta$ splitting is 300 MeV from the spin-spin force with a suitable fine structure constant α_s. Since this is large maybe perturbation is not valid. The required α_s becomes much less when nonperturbative effects are taken into account (see Chapter 6).

Involved Faddeev type calculations were done by *Silvestre-Brac* and *Gignoux* (1985) with part of Breit force and linear confinement nonperturbatively. The non-relativistic model, however, is crude.

In their earlier work Isgur and Karl had treated the positive and negative parity states in isolation. The oscillator parameter was taken different to fit spectrum in the two cases. Finally *Capstick* and *Isgur* (1986) took 6 shells of oscillators and included relativistic kinetic energy corrections. The Roper resonance (1440, $1^+/2$) state comes 100 MeV too high. Of course if one treats the positive parity states in isolation there is no problem fitting the spectrum and this is evident from the work of Isgur and Karl. Nevertheless this has been again looked at by *Kalman* and *Tran* (1989). They also discuss this problem of doing the odd parity states but do not, in this work, calculate them thus avoiding the problem.

There is some evidence that there may be deformation of the mean field for the excited states (*Murthy* et al. 1984). This is also reviewed in *Bhaduri* (1988).

The decays from the excited baryons to the nucleon ground state have been looked at by various authors. One particular transition has continued to attract much attention (*Gershtein* and *Dzhikiya* 1981, *Isgur* et al. 1982, *Dey* and *Dey* 1984, *Bourdeau* and *Mukhopadhyay* 1987) as this involves D-state admixture into the nucleon ground state via the tensor force. This is the $\Delta \rightarrow N\gamma$ transition. *Bienkowska* et al. (1987) considered the transition in a relativistic model. There was a very general work on the transitions done by *Koniuk* and *Isgur* (1980), where they considered γ decays and the decays to $N\pi$, in the non-relativistic oscillator model. Two of the interesting photon decays come out in better agreement with experiment in the deformed oscillator model (*Murthy* and *Bhaduri*, 1985).

An interesting possibility is to look at the Roper and its analogous states in the strange baryons as bandheads of some $SU(3)$ interacting bosonic states (*Toki, Dey* and *Dey* 1983). Or one may take a super-symmetric extension: a graded Lie Group $SU(15/30)$ worked out by *Zu-Rong Yu* (1988). In this case one puts mesons as well as baryons (*Dey* and *Dey* 1990). It is also possible to include positive as well as negative parity baryons using s, p, d, f and g boson-fermions. In this connection we must mention the early work of *Barut* (1964, 1968) where he had looked at parity-doubled states in the conformal O(4, 2) model. This has been revived in the work of *Freund* and *Rosner* (1992, FR in short); *Dey, Dey* and *Tomio* (1992); *Cudell* and *Dienes* (1992). This is based on old wisdom of the Regge model giving same trajectory for mesons and baryons. Modern version is the conformal string theory of *Kutasov* and *Seiberg* (1990). We shall come back to these papers in the next section.

2.6   The Necessity of Relativistic Models

The non-relativistic models were very successful and no one denies that. Instead one tries to understand *why* they are so successful. The problem is that light u, d and even s quarks move fast, so they should be described by relativistic equations. For example *Bhaduri, Cohler* and *Nogami* (1980) had looked at velocities of *constituent* quarks and found them to be about $1/2$ of the velocity of light! Of course current quark masses are much smaller, about 10 MeV for u, d and about 150 MeV for s. They have even larger velocities, one cannot indeed *justify* non-relativistic physics.

The question of a Schrödinger versus Dirac equation for a massless quark was considered by *Bhaduri* and *Brack* (1982). They showed that spectrum produced by massless quarks in a Dirac equation with linear potential is essentially the same as that of 500 MeV quarks in the Schrödinger picture with a shallower confinement. The Dirac magnetic moments for the excited states decrease rapidly. This can be simulated in the Schrödinger case by adding an extra term to the mass given by the equivalent energy difference of the level concerned with the ground state.

The classic model of relativistic quarks and gluons is to confine them in the so-called MIT bag by *Chodos* et al. (1974). The model is simple and there are problems with the excited states and also with centre of mass subtraction. But because of the simplicity of the model it has survived and every now and then we turn to it to do simple calculations. The model assumes that the quarks are confined in the bag because of a pressure difference B, between the perturbative vacuum inside the bag and the QCD vacuum outside. The large and small components match on the bag boundary. A spherical bag is assumed in most cases and then the wavefunctions are just spherical Bessel functions. Of course there is the large and the small component of the wave functions and the phenomenology therefore is richer compared to that of the non-relativistic case. One can write the energy of the hadron in the model as

$$M = nx/R + BV + Z/R \qquad (2.6.1)$$

where the first term is the quark kinetic energy (for baryon $n = 3$, and $n = 2$ for meson). To obtain x it is enough to use the boundary conditions at a radius R. BV is the volume energy needed to dig a hole in the QCD vacuum. The third term contains centre of mass correction, one-gluon-exchange contribution and might also contain a Casimir energy term which must be present as soon as we talk of two vacua. This Casimir energy is the difference of the infinite fluctuation energies of the gluon and quark fields. This is a finite number even though the two vacua have infinite energies which are to be renormalized. Obviously the calculation of the difference of two such quantities are ambiguous and the first calculation for Abelian gluons was done by Boyer. This was followed by many workers including *Schwinger* et al. (1979), and *Milton* (1983). The quark Casimir energy is much smaller than the Abelian gluon case and was calculated by *Milton* (1983). There are no calculations for non-Abelian gluons. The centre of mass correction in the last term of Eq. (1) is somewhat ad hoc. Indeed one can fit the factor x, Z and B to the known ground state baryons and this was done by *Aerts* and *Rafelski* (1984) ten years after the original paper. The Eq. (1) enables one to calculate the equilibrium radius. For this equilibrium one gets

$$M = 4BV \qquad (2.6.2)$$

and the pressure is $1/3$ the energy density as for radiation. Indeed for massless relativistic particles this is what one expects.

One can take the second derivative of the energy (Eq. (2)) and one finds that the compressibility K of the system is equal to its energy. What is interesting is that one can show that K is equal to that of the non-relativistic models. From this the breathing mode type of excitation can be estimated and this comes out too high, about 800 MeV. So one cannot identify the Roper (1440) as the breathing mode. These authors also related the compressibility of the nucleon to that of nuclear matter extrapolated to higher density. This was done using Nuclear Skyrme interaction and another appropriate interaction called $G - 1$ (see *Bhaduri, Dey* and *Preston* 1984 for details). Also one can look at quark-gluon gas and get its K. All three K-s seem to overlap around 1.5 to 2 times normal nuclear density! Of course in each case the interactions may be questioned and the extrapolation from nuclear matter should not be taken too literally.

The surface of the bag can be dynamic and can itself vibrate. This was exploited to solve, for example, the problem of the low lying Roper (1440) state by *Tomio* and *Nogami* (1985) and *Suzuki* et al. (1985).

The question of excited states in the bag model is a tricky one. It was suggested in the original paper that one can look at the density of states $\rho(m)\mathrm{d}m$ between m and $m + \mathrm{d}m$ and *this* gives us a limiting temperature τ_0 into the bargain. The τ_0 was found from dual resonance bootstrap by Hagedorn and theory further developed by him with *Rafelski* (1980, 1981). The density $\rho(m)$ goes as $\exp(m/\tau_0)$. Laplace transform of the partition function of the states yields $\rho(m)$.

Kapusta (1981 and 1982) worked this out and showed that one can put the centre of mass correction in the Laplace transform as a constraint. Thus one of the major problems of the bag model was solved. He also showed that the partition function can be calculated from the single particle states and some corrections can be calculated due to perturbative gluon exchange interactions.

Jennings and *Bhaduri* (1982) found that discrete states for Abelian gluons in confined volumes produce important corrections to the exponential form of $\rho(m)$. This was done for spherical bags containing quarks by *Bhaduri, Dey* and *Srivastava* (1985). In the last two references a smooth density $g(\varepsilon)$ of single particle energy states, ε, was assumed . It was shown by *Dey, Dey* and *Ghose* (1989) that discrete states cannot be replaced by a smooth density $g(\varepsilon)$ without making serious errors in free energy and other thermodynamic functions, except at a particular temperature T_s. It was also shown that the bag survives upto a temperature T_c beyond which the pressure generated by the quarks and gluons inside the bag exceeds B. However there is a temperature range $T_s \leq T \leq T_c$ where one gets *large bags*.

The density of states $\rho(m)$ for mesons using the model of *Dey, Dey* and *Ghose* (1989) gives very good fit to the observed mesons (*Dey, Tomio* and *Dey*, 1990). The work can be extended to baryons introducing a chemical potential to constrain the baryon number as done in *Ansari* et al. (1990), where colour projection was also done. Qualitative results do not change due to colour projection. $\rho(m)$ for baryons is yet to be done. Another interesting approach is provided by *Kutasov* and *Seiberg* (1990). Without going into any detail we quote from them. The appearance of the destabilizing tachyons in a string theory severely constrains the difference of the densities of bosons and fermions in that theory. Their result shows that tachyon elimination does not require full-fledged supersymmetry. Cancellation between the boson and fermion states is all that is needed. It turns out that though the density of states of mesons and baryons each rises exponentially with energy, their difference rises only like a low exponent of energy. It was shown by *Cudell* and *Dienes* (1992) that the meson to baryon ratio in $\rho(m)$ fluctuates around unity in these string models.

As mentioned before, FR have shown that the experimental meson and baryons densities are indeed very close. We show it slightly differently in Fig. 2.1 by placing the mesons and baryons in a histogram with 200 MeV bins. It has been argued in the past that in this way one can produce a spectrum which may be compared with a model in which the u and the d quarks are taken to be degenerate with the s-quark. This is because the mass difference of 150 MeV or so, between the (u, d) and the s, can be washed out on the average, as it were, by the width of the bin. Added to this of course is the fact that the excited states have width of about 50 MeV or more and this also needs to be wrapped up in the bin size.

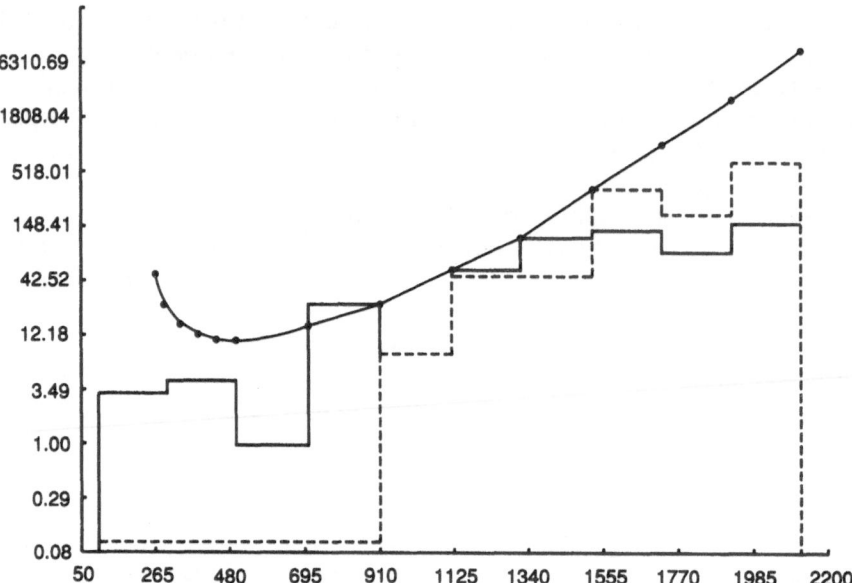

Fig. 2.1. The theoretical bag model density of states $\rho(m)\Delta m$ (curve compared with (1) the mesonic histogram (solid line) and (2) the baryonic one (dashed line). Note that from 0.9 to 1.3 (1) and (2) nearly overlap and the curve fits (1) from 0.7 to 1.3 GeV. Above 1.3 GeV, one need more bosons to equalize with (2) and similarly above 1.7 GeV one probably needs more fermions

Calculating density of states $\rho(m)$, between m and $m + dm$, from the bag is straightforward. We calculated the entropy S of a bag with discrete states and found the stable point with respect of variation with inverse temperature β (*Dey et al.* 1992):

$$S(\beta) = S_o(\beta) + \frac{(\beta - \beta_0)^2}{2} \frac{\partial^2}{\partial \beta^2} S(\beta)|_{\beta_0} + \dots \qquad (2.6.3)$$

The partition function is

$$Z(\beta) = \int_\alpha^\infty \rho(m)e^{-\beta m} dm \equiv e^{\{s(\beta) - \beta m\}}$$

So $\rho(m)$ can be found out to be the Laplace transform of $S(\beta)$:

$$\rho(m) = \frac{1}{2\pi i} \exp\{S(\beta_0)\} \int_{\beta_0 - i\infty}^{\beta_0 + i\infty} \exp\left[\frac{(\beta - \beta_0)^2}{2} \frac{\partial^2}{\partial \beta^2} S(\beta)|_{\beta_0} \right] d\beta$$

We found that the calculated mesonic mass spectrum $\rho(m)$ agrees much better with the experimental mesonic spectrum compared to earlier calculations but the comparison really calls for replacing the meson spectrum by the baryonic one as advocated by FR. Thus if we take the conclusions of FR to be right, namely that the mesonic experimental spectrum is deficient and should be replaced by the baryonic spectrum from about 1.3 GeV onwards, then we get a good fit between theory and experiment over a range of about 1 GeV. Below this one expects the assumption of a continuous distribution $\rho(m)$ to break down. Above this region probably there are exotic baryons just as there may be q^2q^{-2} meson states in the region where experimental $\rho(m)$ mesonic is deficient. One estimates the extra energy needed to create two more quarks (antiquarks) in the bag model as being twice the energy of the bag state, i.e. about $2.04 \times 2\hbar c/R$ which is 1150 MeV for $R = 0.7$ fm. Thus it is the 1.3 GeV region where one might expect q^2q^{-2} meson states in the bag model. Since the baryon radii are higher, one may expect $q^4\bar{q}$ states about 800 MeV above the nucleon at about 1.7 or 1.8 GeV region. Indeed the $S = +1$, Z baryons are reported at 1700, 1725, 1865 MeV etc. One must be cautious: the theoretical estimates and the experimental states need confirmation.

An interesting question one can ask in the statistical model of the hadrons is how many degrees of freedom are involved? This can be calculated in the bag model as well as the non-relativistic model through the specific heat (*Bhaduri* and *Dey*, 1983). This is discussed in detail also in Bhaduri's book (1988).

2.7 Fully Relativistic One-Body Potential Models

One can develop relativistic models with potentials. Immediately one faces the problem of where to put the potential. In the Dirac equation there is the energy and the mass, both linearly present. One can put the potential as fourth component of a vector so that it goes with the energy or put it as a scalar with the mass.

The first successful model was developed by *Leal Ferreira* (1977). For oscillator or linear one-body confining potential he found analytic solutions: relativized oscillators or Airy functions. Later *Leal Ferreira* and *Zagury* (1977) discussed the oscillator confinement in more detail and the detailed linear confinement was done in *Leal Ferreira* et al. (1980).

Palladino and *Leal Ferreira* (1989) have completed a work which probably pushes this model to its ultimate limit. Major questions that arise in a model of this kind are: (1) Can one describe the mesons and baryons with the same sort of interaction? (2) What is the quark mass one has to use to fit the baryons and mesons? In a relativistic formalism for (u, d) quarks one must use masses as small as 5–10 MeV. Often enough one is forced to use larger masses. It is also justified to use running quark masses. In the work of Palladino and Leal Ferreira they do indeed fit the mesons and baryons consistently with (u, d) masses (5, 9) MeV and they include meson loops for the baryons (with $f^2_{NN\pi} = 0.08$ for example) as well as the mesons. The baryon radius comes out to be about 1 fm whereas the meson radii are about 0.38 fm. It will be interesting to see how the model fits the excited mesons [for example the f_0 (975) believed to be a $L = 1$ meson or the pion radial excitation $\pi(1300)$] or the excited baryons. Already the model has an advantage: with equal (vector + scalar) form of the potential the effective spin-orbit splitting cancels (for example between the $p_{3/2}$ and $p_{1/2}$ and excited nucleon states which are experimentally almost degenerate). As we shall see later in more complicated models for the mesons (*Crater* and *van Alstine*, 1984) and baryons (*Dey, Dey* and *Le Tourneux*, 1986) this prescription was adopted for two body potentials. According to Crater and van Alstine for right ordering of mesons this is essential. Probably for the odd parity states of the nucleon also this would be important. Incidentally Crater and van Alstine took large quark masses of 258 MeV for (u, d) quarks.

This brings us on to other calculations where relativistic one body potential models were taken where the potential is scalar. One example is the cubic potential used in the work of the Regensburg group (*Tegen* et al. 1982). Calculation of corresponding meson properties were not done for these potentials. Also it would be interesting to see if some problem arises when excited mesons and baryons are calculated from the strong spin-orbit potential these models imply.

The approach of *Bethe Salpeter* (1951, *BS* in short) has not led to a real breakthrough in our understanding of quark-quark forces. The reason is twofold: first we have no method for comparing the kernel of the BS equation. Second, even if we knew the kernel, we would be unable to solve the equation. If the BS equation cannot be solved and even the non-relativistic limit is delicate as it is, why not start from some other type of relativistic wave equation that can be solved exactly or numerically? The answer to this question is discouraging, according to a recent review team (*Lucha* et al. 1992). There are so many suggestions for relativistic wave equations, semirelativistic treatment, etc. in the literature that any attempt to classify and judge this material appears hopeless; it appears unlikely that any body will ever take the enormous and thankless trouble to review all this work. We will therefore review only one set of calculation, our own work, which has two advantages: (1) asymptotic freedom is built in through the *Richardson* (1979) two body potential, (2) direct application to nuclear matter can be done. This is based on large N_c justification for mean field and uses meson-sector potentials to fit baryons.

2.8 Relativistic Hartree-Fock Models

Although QCD was formulated for three colours, many of its features are more readily understandable if one lets the number of colours N_c become arbitrarily large. This suggestion of *'t Hooft* (1974) became an extremely powerful tool for understanding baryons after the work of *Witten* (1979). $(1/N_c)$ expansion gives, to leading order, most of the qualitative features expected of low energy QCD, such as chiral-symmetry breaking (*Coleman* and *Witten*, 1980) and even confinement in some cases (*Witten*, 1979). For mesons it leads to a Bethe-Salpeter equation with a $q\bar{q}$ potential, while for baryons it yields a Hartree-Fock type equation. An alternative for baryons is to solve classical meson Lagrangians, as in the Skyrme model (*Skyrme*, 1961). And in this model one can put the other mesons with: (1) hidden symmetry (see *Bando* et al. 1988 for a review of their pioneering work, Novozhilov for a recent contribution 1990) or (2) more generally as in the work of the Paris-Orsay group (*Lacombe* et al. 1988). On the other hand one may choose to *keep* the quarks instead of integrating them out of the QCD action in favour of the mesons. This was done by Witten for $(2 + 1)$-dimension with interesting results and also for the non-relativistic heavy quarks.

Starting from the action for a system of interacting quarks and gluons one can obtain, after a series of approximations, a Dirac Hamiltonian with a two body static potential. It was shown by 't Hooft that such a classical (as opposed to field theoretic where $q\bar{q}$ loops essentially introduce infinite degrees of freedom) two body interaction may be derived by summing all the gluon loops that can be drawn on a plane. Witten further showed that this interaction, which is appropriate for the essentially two body meson, can also be used in the mean field approximation for baryon in the same order. Present day techniques do not permit summing up all the planar gluon diagrams which would yield such a potential unambiguously. As an alternative one can borrow a potential from the meson calculation, for example that of Crater and van Alstine mentioned before and test it for a baryon (*Dey, Dey* and *Le Tourneux*). The potential used in this case is due to Richardson and it passes the test very well. We will describe the calculation in some detail, as there is the possibility of extending such calculations and there are other potentials with similar properties as Richardson.[1]

The Hamiltonian reduced to

$$H = \sum_{i=1}^{N} (\alpha_i p_i + m_i \beta_i) + \frac{1}{4} \sum_{i<j} \lambda(i) \cdot \lambda(j) V(r_{ij}) \tag{2.8.1}$$

where the two body potential $V(r_{ij})$ is the phenomenological $q\bar{q}$ potential. Assuming that all N_c quarks occupy the same state $\Phi(r)$, we have the Slater determinant for the baryon ground state

$$\Psi = \Pi \alpha^{+}_{fcjm} |0\rangle \tag{2.8.2}$$

where α creates a quark of colour c flavour f in the state (jm). Since

$$\langle \text{colour singlet} | \lambda(i)\lambda(j) | \text{colour singlet} \rangle = -2(N_c + 1)/N_c \tag{2.8.3}$$

the energy average is

$$E = N_c \int \Phi^{+}_{jm}(r) \, (\alpha \cdot p + \beta m)\Phi_{jm}(r) dr$$

$$- (N_c^2 - 1)/4 \int \Phi^{+}_{jm}(r_1)\Phi^{+}_{jm}(r_2)V(r_{12})\Phi_{jm}(r_1)\Phi_{jm}(r_2) dr_1 dr_2 \tag{2.8.4}$$

Varying this with respect to Φ along with the orthogonality condition we obtain the HF equation

$$\left[\alpha \cdot p + \beta m - (N_c^2 - 1)/2N_c \int \Phi^{+}_{jm}(x)V(r - x)\Phi_{jm}(x)dx \right] \Phi_{jm}(r)$$

$$= \epsilon \, \Phi_{jm}(r) \tag{2.8.5}$$

---

[1] Incidentally there are also other attempts at solving the relativistic meson problem apart from the Dirac constraint method used by Crater and van Alstine (see for example *Leal Ferreira*, 1989).

where ϵ is the single particle energy. The second term in the above equation is obviously the single particle potential or the mean field obtained self consistently

$$\omega_{av}(\underline{r}) = -\frac{N_c^{-1}}{2N_c} \int \Phi_{jm}^+(\underline{r}') V(\underline{r} - \underline{r}') \Phi_{jm}(\underline{r}') d\underline{r}' \tag{2.8.6}$$

and is local. From the derivation it is clear that the mean field is a vector one. The Hartree-Fock energy E is given by

$$E_{HF} = N \left(\epsilon - \frac{1}{2} \left\langle \Phi_{jm} | \omega_{av} | \Phi_{jm} \right\rangle \right) \tag{2.8.7}$$

where ϵ is the single particle energy obtained by solving

$$\left[\alpha \cdot p + \beta m + w_{av}(\underline{r}) \right] \Phi_{jm}(\underline{r}) = \epsilon \, \Phi_{jm}(\underline{r}) \tag{2.8.8}$$

equations (6–8) are to be solved self-consistently for quarks in the lowest orbital,

$$\Phi_{12\frac{1}{2}m}(\underline{r}) = \left[\frac{1}{4\pi} \right]^{1/2} \left[\begin{array}{c} iG(\underline{r}) \\ (\alpha \cdot \underline{r})F(\underline{r}) \end{array} \right] \chi_m \tag{2.8.9}$$

and the Eq. (8) yields coupled differential equations:

$$\frac{dG}{dr} - (m - w_{av} + \epsilon)F = 0$$

$$\frac{dF}{dr} - \frac{2}{r}F + (\epsilon - w_{av} - m)G = 0 \tag{2.8.10}$$

Thus starting with an initial wave function the single particle potential w is derived iteratively.

But one difficulty is that the Hartree Fock solutions violate translational invariance, since they are formed of single particle wave functions derived from an average potential. As a consequence, the centre of mass motion is not well defined in Hartree Fock solutions and this entails a spurious contribution from the centre of mass kinetic energy to the total energy. Since the relative importance of this spurious effect increases as the number of particle decreases, it is important that it should be corrected for systems formed of few particles. This can be done by extending to the relativistic HF equations the Peierls-Yoccoz procedure of nuclear physics.

When one looks at the nucleon and isobar with u, d quarks of negligible masses ($m = 10$ MeV) there is a problem. The particles are not confined even though the Richardson potential contains a linearly rising confining part. This is due to the fact that the effective single particle confining potential is a vector one. The same problem was encountered by Crater and Van Alstine who suggested the prescription of taking a half-vector half-scalar form for the linear part. This choice also leads to a cancellation of spin orbit effects at long range, thus preventing partial multiplet inversions for the lighter mesons. The same procedure was adopted by Dey, Dey and Le Tourneux and reasonable results were obtained for the isobar after subtraction of the centre of mass kinetic energy. The half-vector and half-scalar form implies one-body potentials, a vector $U(r)$ and a mass $m(r)$, both growing with distance. For example $m(r)$ reaches a value of 200 Mev around $r = 2$ fm. and continues to grow.

We may take the scale of SBCS much smaller than that of confinement (*Shuryak*, 1988). This was modelled in *Dey* et al. (1990a).

The problem is that one cannot use a co-ordinate dependent form for the $m(x)$ in a Lagrangian formalism – but has to introduce a field $\sigma(x)$ instead. This is an effective field and may therefore change with the ambient density. This relates the quark problem to the nuclear matter problem and it is possible to find consistent solutions for the nuclear binding energy curve from this. One has to invoke a vector ω field also and as stated by *Guichon* (1988) and *Frederico* et al. (1989), who first suggested this kind of treatment, – these are not fundamental at the quark level. One finds (*Dey* et al. 1990b) that if one couples this approach with the relativistic HF at the quark level, the stringent condition of matching the $\sigma(r)$ with the quark $\bar{\psi}\psi$ determines the sigma potential, $U(\sigma)$ and mass of the sigma scales as f_π with density. $U(\sigma)$ looks like that found in soliton bag models. One can fit nuclear matter satisfactorily if one takes an ω-mass which also scales with density like f_π and takes a standard form for the contribution from the one-pion-exchange potential to nuclear matter (given by for example *Cenni* et al. 1985). This latter contribution from the O.P.E.P. is quite justified, since QCD sum rules predict essentially the correct value for the pion-nucleon coupling constant (*Reinders, Rubinstein* and *Yazaki*, 1982, 1985) and this coupling constant has very little density dependence around normal nuclear matter density (*Dey, Dey* and *Ghose*, 1985). The scaling of the mass of ω with density was expected by *Brown* and *Rho* (1989) from the missing strength of the isoscalar sum rule. It is interesting to find the same behaviour is *essential* to get a fit to nuclear matter starting from a quark mean field model.

3 Currents, Anomaly, Solitons and Fractional Fermions

3.1 The Theorem of Emmy Noether

Let us start with a Lagrangian density $\mathcal{L} = \mathcal{L}(\varphi, \partial_\mu, \varphi)$ where the φ is a scalar field. We write φ_μ for the space-time derivative. We may have a set of independent fields φ^i and their derivatives φ^i_μ. where i runs over the field indices when there are more than one field. Sometimes we will drop this index i.

One can define the generalized momentum:

$$\Pi^\mu_i = \partial\mathcal{L}/\partial\varphi^i_\mu \tag{3.1.1}$$

Now use the equal time canonical commutation relations

$$i[\Pi^0_i(t, \underline{x}),\ \varphi_j(t, \underline{y})] = \delta(\underline{x} - \underline{y})\delta_{ij} \tag{3.1.2}$$

The commutator of φ with itself and Π with itself vanishes. The Euler–Lagrange equation of motion is given by

$$\partial_\mu \Pi^\mu = \partial\mathcal{L}/\partial\varphi \tag{3.1.3}$$

Let us envisage a change whereby φ goes over to $\varphi + \delta\varphi$, and consequently $\mathcal{L}$ to $\mathcal{L} + \delta\mathcal{L}$. We say there is a symmetry if the $\mathcal{L}$, integrated over space-time, in other words the action, is invariant under such transformations. It is possible then to write $\delta\mathcal{L} = \partial_\mu\Lambda^\mu$, since this can be transformed into a surface integral and will vanish for a well behaved Λ. This Λ should be found out without using the equations of motion. Once it is known, one can then get an expression for a conserved current as follows:

$$\delta\mathcal{L} = \frac{\partial\mathcal{L}}{\partial\varphi}\delta\varphi + \frac{\partial\mathcal{L}}{\partial\varphi^\mu}\partial^\mu(\delta\varphi) = \partial_\mu\Lambda^\mu \tag{3.1.4a}$$

$$= (\partial_\mu\Pi^\mu)\delta\varphi + \Pi_\mu(\partial^\mu\delta\varphi) = \partial_\mu(\Pi^\mu\delta\varphi) \tag{3.1.4b}$$

Note that only now, in Eq. (4b) we have used the equation of motion (3), and also the definition of momentum Eq. (1) to get Eq. (4b) in the form of a total derivative, $\partial_\mu(\Pi^\mu\delta\varphi)$. The current

$$j^\mu = \Pi^\mu\delta\varphi - \Lambda^\mu \tag{3.1.5}$$

is then conserved:

$$\partial_\mu j^\mu = 0 \tag{3.1.6}$$

This gives rise to a conserved (time independent) charge

$$Q = \int j^0 \mathrm{d}^3 x \qquad (3.1.7)$$

This is seen by integrating Eq. (6) over 3-volume:

$$\int \partial_0 j^0 \mathrm{d}^3 x + \int \partial_m j^m \mathrm{d}^3 x = 0 \qquad (3.1.8)$$

Second term vanishes after being transformed into a surface integral and integrated over the surface taken far away. Thus

$$\frac{\partial}{\partial t} \int j^0 \mathrm{d}^3 x = \frac{\partial Q}{\partial t} = 0 \qquad (3.1.9)$$

This is Noether's theorem. Formally stated, the theorem is:- if there is a symmetry of the Lagrangian, there is a conserved current and the associated charge is independent of time. The corresponding current is called Noether current after her.

3.2 Internal Symmetry and Space-Time Symmetry

If $\Lambda^\mu = 0$ we call the symmetry "internal". If Λ^μ is non-zero it is a space-time symmetry. This is a loose statement, – one should be a bit more cautious with these definitions. The problem is that one can add any total derivative $\partial_\mu X^\mu$ to $\mathcal{L}$ and provided the function X is well behaved at infinity the two definitions lead to an identical action. One can replace $\delta\mathcal{L}$ by

$$\partial \mathcal{L}' = \delta \mathcal{L} + \partial_\mu \delta X^\mu = \partial_\mu (\Lambda^\mu + \delta X^\mu) \qquad (3.2.1)$$

If it is not possible to find such a new $\mathcal{L}' = \mathcal{L} + \delta\mathcal{L}$ then only do we have a space-time symmetry. We will illustrate these by simple examples.

3.3 Standard Examples

(i) Flavour SU(3) of Gell-Mann and Ne'eman

Nuclear physicists are familiar with isospin symmetry – i.e. the $SU(2)$ symmetry between the neutron and the proton. This was generalized to include strange particles, introducing strangeness or hypercharge to extend the symmetry to flavour $SU(3)$. In quark language this implies assuming a symmetry between the u, d and the strange quark s. This symmetry is not exact, but is broken since the s-quark is heavier by about 150 MeV. We have

$$\delta^a \varphi = T^a \varphi, \qquad a = 1, 2, \dots, 8 \qquad (3.3.1)$$

where T^a are the generators of the internal symmetry group. So from Eq. (3.1.5) we find a conserved current:

$$j_\mu^a = \Pi_\mu T^a \varphi \qquad (3.3.2)$$

(ii) Translation as an Example of Space-Time Symmetry

Under a translation we have

$$\mathcal{L}(x + a) = \mathcal{L}[\varphi(x + a), \; \varphi_\mu(x + a)] \qquad (3.3.3)$$

and

$$\delta\varphi = \delta a^\mu(x)\varphi_\mu(x) \qquad (3.3.4a)$$

$$\delta\varphi_\mu(x) = \delta a^\nu(x)\partial_\mu\varphi_\nu(x) + \partial_\mu[\delta a^\nu(x)]\varphi_\nu(x) \qquad (3.3.4b)$$

After an integration by parts the corresponding variation of the action **I** is

$$\delta\mathbf{I} = \int d^4x \left\{ \partial_\nu\mathcal{L} - \partial_\mu\left[\frac{\partial\mathcal{L}}{\partial\varphi_\mu}\varphi_\nu(x) \right] \right\} \delta a^\nu(x) \qquad (3.3.5a)$$

The conserved quantity is the energy momentum tensor

$$\Theta^{\mu\nu} = \Pi^\mu\partial^\nu\varphi - g^{\mu\nu}\mathcal{L} \qquad (3.3.5b)$$

Here $\Lambda^{\mu\nu} = g^{\mu\nu}\mathcal{L}$.

3.4 Current Algebra and Anomaly

(i) Introduction to Current Algebra

Let us consider internal transformations defined by the Eq. (3.3.1). Even if they are not the symmetry of $\mathcal{L}(\delta\mathcal{L} \neq 0)$ we can still define the current $j_\mu^a = \Pi_\mu T^a \varphi$ as obvious from Eq. (3.1.4) and the charges Q^a. Without any detailed knowledge of $\mathcal{L}$ we can calculate the current commutators determined from canonical commutation relations.

$$[j_0^a(t, \underline{x}), j_0^b(t, \underline{y})] = [\Pi_0(t, \underline{x})T^a\varphi(t, \underline{x}), \; \Pi_0(t, \underline{y})T^b\varphi(t, \underline{y})] \qquad (3.4.1)$$

$$= \Pi_0(t, \underline{x})T^a[\varphi(t, \underline{x}), \; \Pi_0(t, \underline{y})T^b\varphi(t, \underline{y})]$$

$$+ [\Pi_0(t, \underline{x}), \; \Pi_0(t, \underline{y})T^b\varphi(t, \underline{y})]T^a\varphi(t, \underline{x})$$

$$= i\Pi_0(t, \underline{x})[T^a, T^b]\varphi(t, \underline{x})\delta(\underline{x} - \underline{y}). \qquad (3.4.2)$$

Use $[T^a, T^b] = if_{abc}T^c$ where f_{abc} are the $SU(3)$ structure constants. (The analogous relation in $SU(2)$ is $[\sigma^a, \sigma^b] = i\varepsilon_{abc}\sigma^c)$. Then we get

$$[j_0^a(t, \underline{x}), \; j_0^b(t, \underline{y})] = -f_{abc}j_0^c(t, \underline{y})\delta(\underline{x} - \underline{y}) \qquad (3.4.3)$$

So the fourth components of the currents, which are the charge densities, have canonical commutation relations. Similarly we can write for the charges:

$$i[Q^a(t), \varphi(t, \underline{x})] = i \int d^3y [j_0^a(t, \underline{y}), \ \varphi(t, \underline{x})] = T^a \varphi(t, \underline{x})$$

$$= \delta^a \varphi(t, \underline{x}) \tag{3.4.4}$$

They are the generators for the transformations and in the case where the current is conserved, they are Lorentz scalars.

So far we have been cavalier about multiplying two operators (actually they are distribution functions) at the same space-time point. Also, we did not bother whether an equal time limit of an unequal time commutator really exists. These details have far reaching consequences. For example, Eq. (3.4.3) holds good in all Lorentz frames. When it is integrated once we get

$$[Q^a, j_0^b(0)] = -f_{abc} j_0^c(0) \tag{3.4.5}$$

Assuming the current to be conserved and applying an infinitesimal Lorentz transformation to Eq. (3.4.5) we have

$$[Q^a, j_i^b(0)] = -f_{abc} j_i^c(0) \tag{3.4.6}$$

In terms of charge density

$$[j_0^a(t, \underline{x}), \ j_i^b(t, \underline{y})] = -f_{abc} j_i^c(t, \underline{y})\delta(\underline{x} - \underline{y}) + S_{ij}^{ab}(t, \underline{y})\partial^j \delta(\underline{x} - \underline{y})$$

$$+ \text{ higher derivatives of } \delta \text{ function} \tag{3.4.7}$$

where the S_{ij} terms are surface terms which vanish on integration. That these terms occur as a consequence of covariance and positivity was first noticed by Schwinger and these are named after him. We refer the readers to *Itzykson* and *Zuber* (1980, Section 11.3, p. 520) where the Schwinger term is related to the difference between the naive time ordered product and the conserved covariant version, and it is shown how the local commutation rules may be tested. In the next section we shall mention how this is related to the anomaly.

(ii) Introduction to Anomaly

It was demonstrated by Bell and Jackiw and also by Adler that even though for massless quantum electrodynamics the vector and axial currents are conserved classically, it is not possible to conserve both of them in the quantum version. The conservation of the vector and axial vector currents are the Ward identities in momentum space. The violation of the Ward identities is known as the anomaly although there is nothing anomalous about it – it is expected in quantum field theory. In fact the experimentally observed π^0 to 2γ decay is explained nicely by this anomaly. We shall not go into great details about it but this has been derived in many ways and we shall outline some of these *different ways* in the next chapter, as these particular derivations teach us something new. A very nice review is given by *Jackiw* (1985) where he states: "we are not yet at the end of the physics nor of the mathematics that can emerge from understanding anomalies". We list the *different ways* here:

(1) Naive time ordered product of two equal time currents are generally not covariant. The undesirable non-covariant part may cancel with the Schwinger term. This was conjectured by Feynman. Generally this is valid. But in some cases it is not and then Ward identities are not obeyed. One such case is $\pi^0 \rightarrow 2\gamma$. The axial current is chosen to carry the anomaly keeping the vector Ward identity intact.

(2) The same can be found by calculating Feynman graphs directly in momentum space when there are some momentum routing ambiguities. We shall very briefly touch upon this in the next chapter.

(3) The third way is to go to Feynman path integral technique when one can show that the fermion measure is not chirally invariant. We shall discuss this in some detail as in non-Abelian gauge theories this is the general way of showing the anomaly.

(4) The most difficult method concerns the fact that we have now mathematical reasons for understanding that the functional determinant of chiral fermions coupled to non-Abelian fields cannot be defined gauge invariantly.

(5) *Schwinger* (1951) found fluctuations through a modified propagator in the presence of a field and this also gives the correct anomaly. This seems a very promising approach for large fields. Such fields arise in heavy ion reactions when one uses Uranium or Thorium projectiles on Uranium-like targets as in the GSI accelerator in Darmstadt, in Germany. This leads to strong fields which could generate new phases of QED as suggested by Caldi and Chodos. For applications as well as earlier references to the experimental and theoretical work see the beautiful paper by *Accetta*, *Caldi* and *Chodos* (1989) where they seek to relate astrophysical γ ray bursts and the GSI particle.

In the remaining part of this chapter we briefly review fractional fermions and solitons since these are also new physical concepts which will become relevant in Chapter 5.

3.5 Fractional Fermions

Quantum mechanical semi-classical techniques were applied to quantum field theory by *Skyrme* (1961, 1962) but for some years no attention was paid to this idea. But by the mid-1970's we learned how to fit such classical, finite energy configurations – now called solitons – into the fully quantized field theory. The impetus did not come from Skyrme's baryons but rather from vortices and monopoles. Today, the subject is vigorously investigated with emphasis on monopole physics in GUTs (grand unified theories) with a revival of Skyrme's ideas in baryon phenomenology. We will be concerned with the latter and before that with another classical field configuration, this time in a hypothetical Euclidean space-time, namely instantons. But before we do this we will take up the exciting case of fractional fermions – which are solutions for Dirac-like equations in the presence of soliton fields. To do this we need to review solitons briefly.

3.6   A Brief Review of Solitons: Kinks

Let us consider the Lagrangian density of a scalar field φ in $(1 + 1)$ dimension:

$$\mathcal{L} = 1/2(d\varphi/dt)^2 - 1/2(d\varphi/dx)^2 - U(\varphi), \tag{3.6.1}$$

with

$$U(\varphi) = \lambda/4(\mu^2/\lambda - \varphi^2)^2. \tag{3.6.2}$$

There is a $\lambda\varphi^4$ self coupling. Note the mass (φ^2) term with opposite sign (tachyon!). The degenerate vacua of the theory are

$$\varphi = \pm\mu/\sqrt{\lambda} \tag{3.6.3}$$

with the field equation:

$$d^2\varphi/dt^2 - d^2\varphi/dx^2 = -dU/d\varphi = \lambda(\mu^2/\lambda - \varphi^2)\varphi. \tag{3.6.4}$$

Investigating the static solution $\varphi(x, t) = \varphi(x)$, we have

$$1/2(d\varphi/dx)^2 = U \tag{3.6.5}$$

at the vacuum $U(\varphi) = 0$ and $d\varphi/dx \to 0$, so that the integration constant becomes zero.

$$\int d\varphi/\sqrt{U(\varphi)} = \sqrt{2}x + c' \tag{3.6.6}$$

leading to

$$\varphi(x) = \pm\sqrt{\mu^2/\lambda}\,\tanh(\mu x/\sqrt{2}), \tag{3.6.7}$$

$\pm$ in the above corresponds to the kink and antikink solution respectively. The energy density, using Eq. (6), becomes

$$\mathcal{H}(x) = 1/2(d\varphi/dx)^2 + U(\varphi) = 2U(\varphi)$$

$$= (\mu^4/2\lambda)\,\text{sech}^4(\mu x/2) \tag{3.6.8}$$

and the energy upon integration:

$$E = 4\mu^3/(3\lambda\sqrt{2}) \tag{3.6.9}$$

Note the non-perturbative character ($E \propto 1/\lambda$).

Another interesting property is that there is a four current $J_\mu = \varepsilon_{\mu\nu}\partial^\nu\varphi$ and $\partial^\mu J_\mu(x) = 0$ by construction. This is easy to see – since on taking the divergence one has the symmetric operator $\partial^\mu\partial^\nu$. Since $\varepsilon_{\mu\nu}$ is antisymmetric in μ and ν the product is zero. This is not a Noether current, since we did not have to invoke any symmetry of the Lagrangian to prove the conservation of this current, – it is a topological current giving rise to the topological index q:

$$q = \int_{-\infty}^{\infty} J_0(x)dx = \int_{-\infty}^{\infty} (d\varphi/dx)dx \tag{3.6.10}$$

$$= \varphi(\infty) - \varphi(-\infty) \neq 0 \tag{3.6.11}$$

only if multiple vacua are connected.

Consider the trivial solution $\varphi = \pm \mu / \sqrt{\lambda}$, for all x. The energy, $E = 0$ and so is q. This is the vacuum solution and is non-topological in nature. But the kink given by Eq. (7) has $q \neq 0$, and it is a topological solution. Even though this state is higher in energy (given by Eq. (9)), it cannot decay into the vacuum solution. The conservation of topological charge, q, prevents the decay.

3.7 Polyacetylene: One Dimensional Dirac-Type Equation

It has been shown that systems like the quasi-one dimensional conductor transpoly-acetylene $(CH)_x$ possesses excitations carrying sharp fractional charges $(\pm e/2)$ per spin orientations. This was predicted by *Jackiw* and *Rebbi* (1976) and *Su* et al. (1979, 1980). These excitations act as solitons separating domains having different degenerate vacua. In salts such as TTF-TCNQ similar excitations are produced with charges $(\pm e/3)$, $(\pm 2e/3)$ and $(\pm 4e/3)$ respectively (*Schrieffer*, 1985). For more details of the theory we refer to Jackiw's Comments in Nuclear and Particle Physics (1984), which we largely follow.

Let us concentrate on $(CH)_x$. A microscopic Hamiltonian of the system has been proposed by *Su, Schrieffer* and *Heeger* (1979, 1980). In the continuum limit, electron transport is given by a Dirac-type Hamiltonian in one dimension

$$H = \sigma_3 p + \sigma_1 \varphi, \tag{3.7.1}$$

where

$$\sigma_3 = \begin{pmatrix} 1 & 0 \\ 0 & -1 \end{pmatrix}, \quad \sigma_1 = \begin{pmatrix} 0 & 1 \\ 1 & 0 \end{pmatrix} \tag{3.7.2}$$

Here $\varphi(x)$ is the phonon field, it measures the displacement of the carbon atom from its equilibrium position. In the SSH model $\varphi(x)$ is determined self-consistently by the phonon's dynamics, and in the lowest vacuum states $\varphi(x)$ takes the uniform values $\pm \mu$, as illustrated in Fig. 3.1.

The corresponding electron spectrum of Eq. (1) shows a gap between the filled valence band (negative energies) and empty conduction band (positive energies). The matrix structure in Eq. (1) does not arise because of spin (there is no spin in one dimension anyway). The SSH Hamiltonian ignores electron-electron interactions and spin is a passive label. The Hamiltonian describes spin up and spin down electrons separately. Rather the two component wavefunctions which are the Dirac eigenmodes, refer to right-moving and left-moving electron with momentum $\pm |p|$. In addition to the ground states, where the phonon field takes a constant value, there exist stable excited states where $\varphi(x)$ assumes a kink shape, interpolating as x passes from $-\infty$ to $+\infty$ between the vacuum configurations $-\mu$ and $+\mu$. This is the soliton, S, and it corresponds to a defect in the alteration pattern, as is

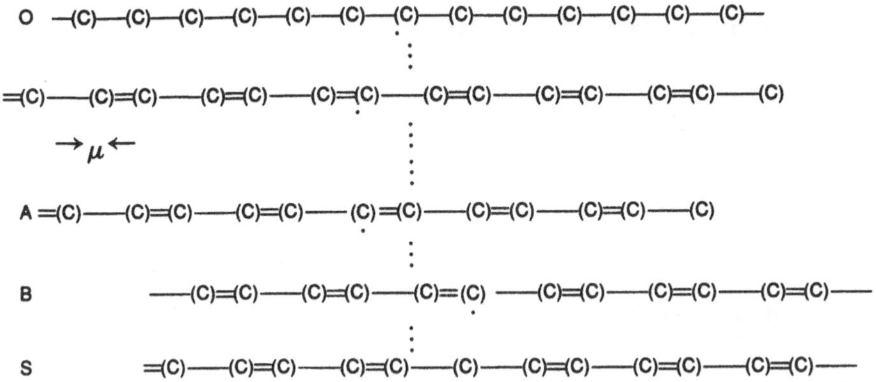

O —(C)—(C)—(C)—(C)—(C)—(C)—(C)—(C)—(C)—(C)—(C)—(C)—(C)—

=(C)——(C)=(C)——(C)=(C)——(C)=(C)——(C)=(C)——(C)=(C)——(C)=(C)——(C)

→ μ ←

A =(C)——(C)=(C)——(C)=(C)——(C)=(C)——(C)=(C)——(C)=(C)——(C)

B   —(C)=(C)——(C)=(C)——(C)=(C)——(C)=(C)——(C)=(C)——(C)=(C)——

S   =(C)——(C)=(C)——(C)=(C)——(C)——(C)=(C)——(C)=(C)——(C)=(C)——

Fig. 3.1. O, A and B are normal polyacetylene chains. The equal spacing configuration is left-right symmetric but has higher energy and is energetically unstable. A and B are the two degenerate vacua breaking the symmetry. S is the soliton chain with a defect in the pattern

exhibited in Fig. 3.1. The Hamiltonian admits a conjugation symmetry:

$$c = \sigma_2 = \begin{pmatrix} 0 & -i \\ i & 0 \end{pmatrix}, \tag{3.7.3}$$

this is ordinary charge conjugation invariance, in the absence of Coulomb inter-action. One easily solves the zero eigenvalue problem and finds that with a kink background a normalizable zero-energy solution exists. Thus our general analysis predicts that fermion number, here coinciding with the electric charge, fractionates to $\pm 1/2$ in the one-soliton state. One should add that fractional charge has not been observed in experiments in polyacetylene. The reason is that electrons come in two species, spin up and spin down. Since both contribute to the number density, the charge in the unfilled state is $(-1/2) \cdot 2 = -1$ while the filled state is neutral $-1 + 1 = 0$. Nevertheless charge fractionalization leaves a spur: the soliton state with unfilled zero energy has net charge, but no net spin since all the electrons are paired. When the level is filled there is no net charge, but now there is net spin. These spin-charge assignments (charged-without-spin, neutral-with-spin) are unexpected, but in fact have been observed and provide experimental verification for the soliton picture of polyacetylene.

4 More on Chiral Anomaly

4.1 Chiral Anomaly in QED

Start with the Dirac-Maxwell Lagrangian density. Let us consider

$$\mathcal{L} = (1/4e^2)F_{\mu\nu}F^{\mu\nu} + \bar{\psi}i\gamma_\mu D^\mu\psi - m\bar{\psi}\psi$$
$$D^\mu = \partial_\mu + A_\mu \tag{4.1.1}$$

The covariant derivative differs from the normal form $\partial_\mu - ieA_\mu$, A_μ absorbs a factor $-ie$. Using the gauge transformation $\psi(x) \Rightarrow e^{i\alpha(x)}\psi(x)$ we find the conserved vector Noether current $J_\mu = \bar{\psi}\gamma_\mu\psi$

$$\partial^\mu J_\mu = 0 \tag{4.1.2}$$

Under the transformation $\bar{\psi}\psi$ is invariant. The extra term in the derivative is canceled by putting $A_\mu \Rightarrow A_\mu - i\partial_\mu\alpha$. This is the well-known gauge transformation in electrodynamics. One can have a transformation $\psi(x) \Rightarrow e^{i\theta\gamma^5}\psi(x)$. The square of γ^5 is one and it anticommutes with the other $\gamma-s$. This is a global transformation and so it is not necessary to change the potential function and for the massless case the current $j_\mu^5 = \bar{\psi}\gamma_\mu\gamma^5\psi$ is conserved. For the massive case it is not, since now $\bar{\psi}\psi$ is not an invariant anymore. However as we shall see in the massless case also the current j_μ^5 is not conserved in quantum field theory. This is the chiral anomaly, but we will come to it a little later.

4.2 Gentleness

When we have $\partial^\mu J_\mu = $ *something non-zero* and the operator on the right hand side is something *gentle*, which means an operator of lower dimension, – we say the current is partially conserved. Consider the Lagrangian density

$$\mathcal{L} = \bar{\psi}i\gamma_\mu\partial^\mu\psi - m\bar{\psi}\psi + i\bar{\psi}\gamma_\mu A^\mu\psi - g\varphi\bar{\psi}\gamma_5\psi \tag{4.2.1}$$

where the kinetic energy of the boson fields are neglected, they are constant external fields, for example. φ is a pseudoscalar field. The equations of motion are

$$i\gamma^\mu\partial_\mu\psi = m\psi - i\gamma^\mu\psi A_\mu + g\varphi\gamma_5\psi, \tag{4.2.2}$$

$$i\partial_\mu\bar{\psi}\gamma^\mu = -m\bar{\psi} + i\bar{\psi}\gamma^\mu A_\mu - g\varphi\bar{\psi}\gamma_5. \tag{4.2.3}$$

Now one can work out the derivative of j_μ^5 to get

$$\partial^\mu j_\mu^5 = 2im\bar\psi\gamma_5\psi + 2ig\varphi\bar\psi\psi. \tag{4.2.4}$$

Let us check the dimensions. $\mathcal{L}$ has dimension $[M]^4$, the fermion wavefunctions are of dimension $[M]^{3/2}$, φ has the dimension of mass, so g is dimensionless. The derivative in (4) is $[M]^4$ as it is proportional to $\bar\psi\gamma_5\psi$ which is $[M]^3$. And we have a gentle operator for the derivative of the axial current. Hence J_μ^5 is partially conserved (PCAC in short) in the absence of pseudoscalar coupling i.e. $g = 0$. Thus PCAC is given by:

$$\partial^\mu j_\mu^5 = 2im\bar\psi\gamma_5\psi \tag{4.2.5}$$

But this is *classical* as opposed to quantum field theoretical. We shall see that to simulate quantum effects we need a φ field, an effective pion-like field. Also one has to find some scale for g. How do we construct such effective term in the Lagrangian so that we can redefine the current and get back our PCAC relation at the "tree level"? Tree level calculations are those in which there are no loops.

Adler and independently *Bell* and *Jackiw* showed in 1969 that it is not possible to have simultaneous conservation of vector and axial currents even in the case of massless fermions. This is a purely quantum effect which can be simulated in a classical treatment by taking the Lagrangian density of QED [Eq. (4.1.1)] and adding $\varepsilon_{\mu\nu\rho\sigma}F^{\mu\nu}F^{\rho\sigma}\varphi/f_\pi$ where φ is a pseudoscalar field. With such additional term in $\mathcal{L}$, axial vector current is:

$$J_\mu^5 = \bar\psi\gamma_\mu\gamma^5\psi - (e^2/8\pi^2)\varepsilon_{\mu\nu\rho\sigma}A^\nu F^{\rho\sigma}. \tag{4.2.6}$$

$$= j_\mu^5 - (e^2/4\pi^2)\tilde{J}_\mu$$

where $\tilde{J}_\mu = \varepsilon_{\mu\nu\rho\sigma}A^\nu\partial^\rho A^\sigma$ $\hspace{2cm}$ (4.2.7)

The corresponding charge is

$$Q^5 = \int (j_0^5 - (e^2/4\pi^2)\tilde{J}_0)d^3x \tag{4.2.8}$$

The current (Eq. (2.6)) has zero derivative, so it is conserved when the fermions are massless, i.e. one keeps the vector current conserved and adds $\tilde{J}_\mu$ so that PCAC is recovered. The anomalous $\tilde{J}_\mu$ is not gauge invariant. The charge, however, is gauge invariant and conserved. Therefore it may be considered to be the generator of global chiral transformations.

In non-Abelian theories like the Glashow-Weinberg-Salam, one has anomalies in the gauge current itself. To preserve gauge invariance one has to postulate *cancellation* of anomalies between different sets of fermions. This is possible because the anomaly is always independent of the fermion mass as can be seen above. In this way one gets light leptons to cancel the anomalies of more massive quark families in electro-weak theory. In fact this fixes colour to be 3.

It was pointed out, however, by *Jackiw* and *Rajaraman* (1985) that it is not essential to cancel anomalies in this way – the anomaly does not contradict unitarity. In the model they used, the Schwinger model, there is no problem with renormalization. It is a super-renormalizable model. The anomaly, by breaking the gauge invariance, may introduce masses so that one may dispense with Higgs particles in the theory. Recently it has been claimed by *Mitra* and *Rajaraman* (1989) that an anomalous quantum theory can be reformulated in a gauge invariant way. This is a very interesting subject which may become more topical in the future.

4.3 Momentum Routing Anomalies

In field theory we are always concerned with Green's functions or correlators. Let us consider the decay width $\Gamma(\pi^0 \Rightarrow 2\gamma) \cong 10$ e.V. The matrix element of the transition for the photon four momenta p and q is

$$M(p, q) = \langle \pi^0, k | \gamma, p; \gamma', q \rangle, \tag{4.3.1}$$

where $k = p + q$. In terms of the polarization vectors ε of the photons one can write

$$M(p, q) = \varepsilon_\mu(q)\varepsilon_\nu(p)T^{\mu\nu}(p, q) \tag{4.3.2}$$

We now want to write expressions for the derivatives of the currents and use the idea of gentleness introduced in the previous section. The pion is a pseudoscalar, $T^{\mu\nu}$ must be a pseudotensor and out of two momenta p and q we construct

$$T^{\mu\nu}(p, q) = \varepsilon^{\mu\nu\rho\sigma} p_\rho q_\sigma T(k^2). \tag{4.3.3}$$

In quantum field theory the above amplitude is given by the vacuum expectation value of two time ordered current correlator, the so-called 2-point function:

$$T_{\mu\nu}(k_1, k_2) = i \int d^4x_1 d^4x_2 \langle 0|T[J_\mu(x_1)J_\nu(x_2)P(0)]|0\rangle$$

$$\exp\{i(k_1 x_1 + k_2 x_2)\}, \tag{4.3.4}$$

which involves two vector currents and the pseudoscalar $P(x) = -\bar{\psi}(x)\gamma_5\psi(x)$. If we evaluate the 3-point function with two vector currents and one axial we get:

$$T_{\mu\nu\lambda}(k_1, k_2) = i \int d^4x_1 d^4x_2 \langle 0|T[J_\mu(x_1)J_\nu(x_2)J_\lambda^5(0)]|0\rangle$$

$$\exp\{i(k_1 x_1 + k_2 x_2)\}, \tag{4.3.5}$$

The derivative in momentum space is a tensor multiplication with the momentum operator. So conservation of the gauge current (i.e. four divergence of the vector current being zero) implies:

$$k_1^\mu T_{\mu\nu\lambda}(k_1, k_2) = k_2^\nu T_{\mu\nu\lambda}(k_1, k_2) = 0. \tag{4.3.6}$$

whereas PCAC (Eq. (2.5)) requires the *gentleness* relation

$$-(k_1 + k_2)^\lambda T_{\mu\nu\lambda}(k_1, k_2) = 2m T_{\mu\nu}(k_1, k_2) \tag{4.3.7}$$

After some algebra one can show that the Ward identities (gauge invariance expressed in momentum space, Eq. (6)) are violated unless one adds an extra term to $T_{\mu\nu\lambda}$ [1]:

$$T_{\mu\nu\lambda} = T_{\mu\nu\lambda} - i(e^2/4\pi^2)\varepsilon_{\mu\nu\lambda\sigma}(k_1 - k_2)^\sigma. \qquad (4.3.8)$$

With this $T_{\mu\nu\lambda}$ gauge invariance is satisfied but it results in non-conservation of J_μ^5 even when $m = 0$:

$$- (k_1 + k_2)^\lambda T_{\mu\nu\lambda}(k_1, k_2) = 2mT_{\mu\nu}(k_1, k_2) + i(e^2/2\pi^2)\varepsilon_{\mu\nu\lambda\sigma}k_1^\lambda k_2^\sigma \qquad (4.3.9a)$$

This leads to the relation

$$\partial^\mu j_\mu^5 = 2im\bar\psi\gamma_5\psi + (e^2/8\pi^2)F_{\mu\nu} * F^{\mu\nu} \qquad (4.3.9b)$$

in terms of the dual tensor:

$$* F_{\rho\lambda} = \frac{1}{2}\varepsilon_{\mu\nu\rho\lambda}F^{\mu\nu} \qquad (4.3.9c)$$

But

$$\frac{1}{2}F_{\mu\nu} * F^{\mu\nu} = \partial^\mu \tilde{J}_\mu \qquad (4.3.10)$$

where $\tilde{J} = \varepsilon_{\mu\nu\rho\sigma}A^\nu\partial^\rho A^\sigma$ already given in Eq. (2.7). Now Eq. (2.5) is satisfied. $\tilde{J}_\mu$ provides with the anomaly given by the Eq. (10). In terms of the electric and magnetic field $\underline{E}$ and $\underline{B}$, it is $\underline{E} \cdot \underline{B}$.

In simple terms can one understand the trouble? The trouble is that the triangle diagram that represents the process under consideration has linearly divergent parts. There are two such diagrams and there is a loop integration over all possible momentum in each of them. If one shifts one momentum label to match up with the other one, the divergences cancel. This seems to be a perfectly justified procedure but in fact one gets additional contribution from surface terms. In integrals one can shift variables only if the integrals are non-divergent or even if they are logarithmically divergent. Let us consider the following integral in one dimension, x, to explain the essential idea. Let us shift the function to $x + a$ and expand in Taylor series.

$$\Delta a = \int_{-\infty}^{\infty} dx[f(x + a) - f(x)]$$

$$= a[f(\infty) - f(-\infty)] + a^2/2[f'(\infty) - f'(-\infty)] + \ldots \qquad (4.3.11)$$

If the integral converges or has log divergence then only Δa is zero. For linear divergence clearly there will be a term proportional to the shift a. If the integral was three dimensional then the derivative term would reduce it to a surface integral. As we see also in later chapters, it is the property of the field on the surface at infinity which is responsible for anomalies.

---

[1] Note that e^2 comes from the colour sum over the quark vertex with the photons. The correction here is in the first order in $\alpha = e^2/4\pi$ (*Itzykson* and *Zuber*, 1980, p. 557).

4.4 Anomalous Ward Identity from Path Integral Approach

We shall next look at the derivation of anomaly from path integral, once again without going into too much detail but stressing the important steps. This is essential since for non-Abelian gauge theories this method is as easy as for QED and we need it for the important Atiyah–Singer index theorem. The method was given in two papers by *Fujikawa* (1979, 1980). The anomaly appears in this scheme as the non-invariance of the fermionic measure in the path integral under a chiral transformation. We define the Feynman amplitude to go from one point $x_1 t_1$ to another point $x_N t_N$ in terms of the action S as follows:

$$\langle x_N t_N | x_1 t_1 \rangle = \sum_{\text{all paths } \alpha} \exp(i S_\alpha / h) = \int_{x_1 t_1}^{x_N t_N} D[x(t)] e^{iS/h} \qquad (4.4.1)$$

where D stands for the measure of the integral over all paths. In field theory with $\mathcal{L} = \mathcal{L}(\varphi_i, \varphi_i^\mu)$ the vacuum to vacuum amplitude is formally written as

$$\langle 0_+ | 0_- \rangle = \int D[\varphi_1] D[\varphi_2] \dots e^{i \int \mathcal{L} d^4 x} \qquad (4.4.2)$$

To calculate Green's functions it is often convenient to couple a source $w(x)$ to the field $\varphi(x)$. We shall take a one-component field for simplicity, the generalization to arbitrary number of fields being straightforward. The generating functional is

$$Z(w) = \langle 0_+ | 0_- \rangle = \int D[\varphi] e^{\int i [\mathcal{L} + w(x)\varphi(x)] d^4 x} \qquad (4.4.3)$$

The N-point Green's function is given by

$$\begin{aligned} G_N(x_1 \dots x_N) &= (-i)^N \delta / \delta w(x_1) \dots \delta / \delta w(x_N) Z(w)|_{w=0} \\ &= \langle 0 | T[\varphi(x_1) \dots \varphi(x_N)] | 0 \rangle, \end{aligned} \qquad (4.4.4)$$

Fujikawa pointed out that the path integral measure of the fermion field in gauge-invariant theories (Abelian and non-Abelian) changes under a chiral transformation. Let us first outline the salient points without going into Grassmann algebra of anti-commuting c-numbers. One starts with the basic Lagrangian density

$$\mathcal{L} = i \bar{\psi} \gamma_\mu D^\mu \psi - m \bar{\psi} \psi, \qquad (4.4.5)$$

the kinetic energy term $F_{\mu\nu} F^{\mu\nu}$ of the gauge field is dropped for the moment as it is not relevant for the argument. The generating functional for $w = 0$ is given by

$$Z = \int D[\psi] D[\bar{\psi}] D[A_\mu] \exp[\int d^4 x \, \bar{\psi} (i\gamma_\mu D^\mu - m)\psi]. \qquad (4.4.6)$$

Let us now introduce a local transformation

$$\psi'(x) = \exp[i\gamma_5 w(x)]\psi(x) \qquad (4.4.7)$$

$$\bar{\psi}'(x) = \bar{\psi}(x) \exp[i\gamma_5 w(x)]. \qquad (4.4.8)$$

Then for the infinitesimal $w(x)$ and using the equation of motion the $\mathcal{L}$ changes to $\mathcal{L}' = \mathcal{L} + \delta\mathcal{L}$ with

$$\delta\mathcal{L} = -2im\bar{\psi}\gamma_5\psi w - \partial_\mu w\bar{\psi}\gamma^\mu\gamma_5\psi. \tag{4.4.9}$$

Assume *incorrectly* that the path integral measure $D[\psi]D[\bar{\psi}]$ remains invariant under the transformations given in (7) and (8). Then the only change in Z coming through δL to first order in w is

$$\exp[i\int d^4x(\mathcal{L}+\delta\mathcal{L}) = (1+i\int d^4x\delta\mathcal{L})\exp[\int d^4x\mathcal{L}]. \tag{4.4.10}$$

But

$$\int d^4x\delta\mathcal{L} = \int d^4x(-2im\bar{\psi}\gamma_5\psi w - \partial_\mu w\bar{\psi}\gamma^\mu\gamma_5\psi)$$

$$= \int d^4xw(x)(-2im\bar{\psi}\gamma_5\psi + \partial_\mu\bar{\psi}\gamma^\mu\gamma_5\psi). \tag{4.4.11}$$

Hence

$$Z' = \int D[\psi]D[\bar{\psi}]D[A_\mu]\{\exp[i\int d^4x\mathcal{L}]\}$$

$$[1+i\int d^4xw(-2im\bar{\psi}\gamma_5\psi + \partial_\mu\bar{\psi}\gamma^\mu\gamma_5\psi)] \tag{4.4.12}$$

The generating functional Z should not depend on the *integration variables* ψ, $\bar{\psi}$ and we may demand $Z = Z'$ for $w = 0$. It follows immediately

$$\partial^\mu J_\mu^5 = 2im\bar{\psi}\gamma_5\psi, \tag{4.4.13}$$

the classical relation without the anomaly term. Before we go to Fujikawa let us ask why the fermionic measure changes:

$$D[\psi]D[\bar{\psi}] \neq D[\psi']D[\bar{\psi}'] \tag{4.4.14}$$

The answer has to do with the negative energy solutions of the Dirac equation. For a zero mass Dirac particle one can define left handed and right handed quarks in a Lorentz invariant way. Second quantization in this scheme implies construction of vacuum filling up the negative energy sea for these left and right handed spinors. In the presence of a gauge field however this kind of separate vacua is not meaningful since the positive and negative energy eigenvalues of the chiral states now behave differently. The left handed states move upward and the right handed states move down as the strength of the interaction increases. As a result the vacuum is not invariant under chirality. This is responsible for the Eq. (14) and therefore the anomaly. Before we proceed further we must learn Grassmann algebra.

4.5 Summary of Grassmann Algebra

The problem of the algebra of anticommuting numbers appears as soon as one is trying to find classical analogs of Fermi fields. Such numbers and the properties of the functions of such numbers were fortunately discovered long ago by a mathematician, Grassmann (see *Rajaraman*, 1982, Chapter 9).

(1) Consider Dirac fields $\psi(x)$. Both $\psi(x)$ and $\bar{\psi}(x)$ $(= \psi^{\dagger}(x)\gamma^0)$ obey $\{\psi(x), \psi(y)\} = \{\bar{\psi}(x), \bar{\psi}(y)\} = 0$ and $\{\psi(\underline{x}, t), \psi^{\dagger}(\underline{y}, t)\} = \delta^3 (\underline{x} - \underline{y})$, the equal time commutation relation. This means $\psi(x)$ and $\bar{\psi}(x)$ are independent, non-trivial operators. Classical analog will be anticommuting functions:

$$\{\psi(x), \ \psi(y)\} = \{\bar{\psi}(x), \ \bar{\psi}(y)\} = \{\psi(x), \ \bar{\psi}(y)\} = 0 \qquad (4.5.1)$$

Note that $\bar{\psi}$ is not the adjoint of ψ. If $\bar{\psi}(x) = \psi^{\dagger}(x)\gamma^0$, $\{\psi, \psi^{\dagger}\}$ is zero only if $\psi = 0$. Here doubling of the number of variables is inevitable.

(2) However $\bar{\psi}$ *transforms* like the adjoint of ψ.

(3) One component Grassmann variable a leads to $a^2 = 0$ because of anticommutation. So the algebra consists of 0, 1 and a only. As we shall see most of the Grassmann relations are simpler than that of ordinary numbers!

(4) Define $\int da f(a)$ in the sense of $\int_{-\infty}^{\infty} f(x) dx$ but remember this is not an area.

(5) $\int da (c_1 f_1 + c_2 f_2) = c_1 \int da f_1 + c_2 \int da f_2$, where the c-s are ordinary c-numbers.

(6) $\int da f(a + b) = \int da f(a)$ where b is arbitrary Grassmann.

The last two properties completely determine the integration apart from a normalization. Define norm as $\int da \, a = 1$. One can expand the function $f(a)$ in Taylor series about zero but there is only one term, since $a^2 = 0$. Therefore we get:

(7) $\int da f(a) = f'(0)$. Thus integration is equivalent to differentiation for functions of Grassmann variables!

(8) Multiple integrations are defined by iterations. For two independent variables $a, \bar{a}$ four combinations can be constructed. $\int da \, d\bar{a} f$ is zero for $f = 1, a, \bar{a}$ but is 1 for $\bar{a}a$. The most important application we need is

$$\int da \, d\bar{a} \exp(\lambda \bar{a}a) = \int da \, d\bar{a}(1 + \lambda \bar{a}a) = \lambda \qquad (4.5.2)$$

(9) Consider two component Grassmann variables, $a_i, \bar{a}_i$; $i = 1, 2$. Let us change variables:

$$b = Ma, \quad \bar{b} = N\bar{a} \qquad (4.5.3)$$

where M and N are 2 by 2 matrices and $b, \bar{b}$ are the new independent Grassmann quantities. We have $b_1 b_2 = (\det M)a_1 a_2$. In order to preserve $\int db_1 db_2 b_1 b_2 = \int da_1 da_2 a_1 a_2$ we must have

$$db_1 db_2 = (\det M)^{-1} da_1 da_2. \qquad (4.5.4)$$

Referring back to Eq. (4) we now have $(\det MN)^{-1} \int d\bar{a} \, da \, e^{-\bar{b}b} = 1$. Since $\det MN = \det N^T M$, defining $N^T M = A$, we get a very useful relation:

$$\int d\bar{a} \, da \, e^{-\bar{a}Aa} = \det A \qquad (4.5.5)$$

4.6 Integration Over Fermi Fields

We introduce two arbitrary complete sets of functions, ψ_r, $\bar{\psi}_r$. They are c-numbers and they are orthonormalized:

$$\int d^4x\,\psi_r\bar{\psi}_s = \delta_{rs} \tag{4.6.1}$$

Expand

$$\psi(x) = \Sigma a_r\psi_r(x) \tag{4.6.2}$$

$$\bar{\psi}(x) = \Sigma \bar{a}_r\bar{\psi}_r(x) \tag{4.6.3}$$

where a_r, $\bar{a}_r$ are Grassmann variables. These are the classical limits of the creation and destruction operators. Define the measure

$$d[\psi]d[\bar{\psi}] = \Pi\, da_r\, d\bar{a}_r \tag{4.6.4}$$

We shall be interested in objects like $d[\psi]d[\bar{\psi}]\exp(-S)$ with $S = -\int d^4x\,\bar{\psi}A\psi$ where A is a linear operator and assumed to be Hermitean. Use ψ_r which are the eigenfunctions of the operator A and assume discrete states:

$$A\psi_r = \lambda_r\psi_r \tag{4.6.5}$$

We get as in Eq. (5.5)

$$S = -\int d^4x \sum \bar{a}_r\psi_r^{\dagger}\lambda_s a_s\psi_s = -\sum_s \lambda_s\bar{a}_s a_s \tag{4.6.6}$$

$$\int d[\psi]d[\bar{\psi}]\exp(-S) = \int \prod_r da_r d\bar{a}_r \exp(\sum \lambda_r\bar{a}_r a_r) = \prod_r \lambda_r$$

$$= \det A \tag{4.6.7}$$

4.7 Chiral Transformation

As in QED (Eq. (4.1.1)) the Lagrangian for $SU(N)$ gauge field coupled with fermions is given by:

$$\mathcal{L} = (1/4g^2)F_{\mu\nu}F^{\mu\nu} + \bar{\psi}i\gamma_\mu D^\mu\psi - m\bar{\psi}\psi \tag{4.7.1}$$

with $D_\mu = \partial_\mu + A_\mu$ and the γ matrices are as given in *Bjorken* and *Drell*, 1965. Defining $F_{\mu\nu} = \sum_a F_{\mu\nu}^a T^a$ and $A_\mu = -ig\sum_a A_\mu^a T^a$ we have the matrix-valued field tensor, T^a are the generators of the $SU(N)$ group.

$$F_{\mu\nu} = \partial_\mu A_\nu - \partial_\nu A_\mu + [A_\mu, A_\nu] \tag{4.7.2}$$

The above equation is continued to Euclidean space-time. The metric is $(-1, -1, -1, -1)$, $x^0 \Rightarrow -ix_4$, $A_0 \Rightarrow iA_4$ and $\gamma^\mu D_\mu = \gamma^\mu(\partial_\mu + A_\mu)$

$$\Rightarrow i\gamma^0 D_4 + \gamma^k D_k \equiv \gamma_4 D_4 + \gamma_k D_k = \mathbb{D} \tag{4.7.3}$$

Let us change the basis from x-representation to n-representation by

$$\psi(x) = \sum a_n \varphi_n(x) = \sum a_n \langle x|n \rangle,$$

$$\bar{\psi}(x) = \sum \bar{a}_n \varphi_n^\dagger(x) = \sum \langle n|x \rangle \bar{a}_n \qquad (4.7.4a)$$

where φ-s form a complete orthonormal set of eigenfunctions of $D\varphi_n(x) = \lambda_n \varphi_n(x)$. φ_n-s have good chirality. Now go back to the chiral transformation and apply it to the Grassmann fields.

$$\psi'(x) = \exp[i\alpha(x)\gamma_5]\psi(x) = \sum a'_n \varphi_n(x) \qquad (4.7.4b)$$

One can write $a'_n = \sum c_{nm} a_m$ where

$$c_{nm} = \int dx \varphi_n^\dagger(x) \exp(i\alpha(x)\gamma_5)\varphi_m(x). \qquad (4.7.4c)$$

We have to find the new measure, the transformation of the volume element, remembering integration with respect to Grassmann variable is equivalent to derivation.

$$\int da'_m = d/da'_m = \sum_n \frac{da_n}{da'_m} \frac{d}{da_n} = \int \sum (c_{mn}^{-1}) da_n$$

Because all da_n anticommute and as in Eq. (5.6) we get

$$\Pi da'_n = [\det(c)]^{-1} \Pi da_n \qquad (4.7.4d)$$

Use infinitesimal transformations $\alpha(x)$, then

$$\det(c_{mn}^{-1}) = [\det(\int d^4x \varphi_n^\dagger(x)(1 + i\alpha(x)\gamma_5)\varphi_m(x)]^{-1}$$

$$= [\det\{\delta_{mn} + i \int d^4x \varphi_n^\dagger(x)\alpha(x)\gamma_5\varphi_m(x)\}]^{-1} \qquad (4.7.5a)$$

There is a useful formula often used in path-integration

$$\det(1 + x) = \exp[(Tr \ln(1 + x)], \qquad (4.7.5b)$$

where $\ln(1 + x) = x - x^2/2 + \ldots \approx x$, if x is infinitesimal.

Combining Eqs. (5a) and (5b), summing over n for the trace:

$$\det(c^{-1}) = \exp[-i \int d^4x \alpha(x) \sum_n \varphi_n^\dagger(x)\gamma_5\varphi_n(x)]. \qquad (4.7.6)$$

We put

$$\sum_n \varphi_n^\dagger(x)\gamma_5\varphi_n(x) = A(x). \qquad (4.7.7)$$

There will be a similar factor due to $\bar{a}_n$. So the change is squared and the new measure of Eq. (6.4) is given as

$$d\mu' = d\mu \exp[-2i \int dx \alpha(x)A(x)], \qquad (4.7.8)$$

where $A(x)$ is the source of the anomaly. It is an ill-defined quantity as we shall show in the next paragraph. Recall

$$\langle x|y\rangle = \delta(x - y) = \sum_n \langle x|n\rangle\langle n|y\rangle = \sum_n \varphi_n(x)\varphi_n^\dagger(y), \qquad (4.7.9)$$

where we have just used the complete set of states n.

$\sum \varphi_n^\dagger(x)\gamma_5\varphi_n(x)$ in Eq. (6) is proportional to

$$Tr\sum_n \gamma_5^{\alpha\beta}\varphi_{n\beta}(x)\varphi_{n\alpha}^\dagger(x) \to Tr\gamma_5^{\alpha\beta}\sum_n \varphi_{n\beta}(x)\varphi_{n\alpha}^\dagger(x) \to Tr\gamma_5^{\alpha\beta}\delta(0)\delta_{\alpha\beta},$$

$$(4.7.10)$$

which is ill-defined. To soften the $\delta(0)$ *Fujikawa* (1980) introduced a cut-off procedure as follows:

$$A(x) = \lim_{M\Rightarrow\infty}\sum \varphi_n^\dagger(x)\gamma_5 \exp[-(\lambda_n/M)^2]\varphi_n(x)$$

$$= \lim_{M\Rightarrow\infty}\lim_{y\Rightarrow x}\sum \varphi_n^\dagger(x)\gamma_5 \exp[-(\mathbb{D}/M)^2]\varphi_n(y)\delta(x - y) \qquad (4.7.11)$$

We write the δ function in momentum space representation:

$$A(x) = \lim_{M\Rightarrow\infty}\lim_{y\Rightarrow x}\int \frac{d^4k}{(2\pi)^4}Tr\gamma_5 \exp[-(\mathbb{D}/M)^2]\exp\{ik(x - y)\} \qquad (4.7.12)$$

Using commutators and anticommutators one can write

$$\mathbb{D}^2 = 1/2\{\gamma_\mu, \gamma_\nu\}D_\mu D_\nu + 1/2[\gamma_\mu, \gamma_\nu]D_\mu D_\nu$$

$$= D_\mu D_\mu + 1/4[\gamma_\mu, \gamma_\nu][D_\mu, D_\nu] \qquad (4.7.13)$$

Now comes the interesting result

$$[D_\mu, D_\nu] = \partial_\mu A_\nu - \partial_\nu A_\mu + [A_\mu, A_\nu] = F_{\mu\nu}.$$

So $\mathbb{D}^2 = D_\mu D_\mu + 1/4[\gamma_\mu, \gamma_\nu]F_{\mu\nu}$.
We also write

$$D_\mu D_\mu = \partial_\mu\partial_\mu + X. \qquad (4.7.14)$$

We need not bother about X since it gives no contribution. We change k to k/M. Two things conspire now to simplify the expression. Only the second order term will contribute. Higher powers are killed by $M \to \infty$. The first term vanishes because trace of γ_5 is zero. The second term (linear) also gives zero since it involves $Tr\gamma_\mu\gamma_\nu\gamma_5$. The relevant second order term gives

$$Tr \gamma_5\gamma_\mu\gamma_\nu\gamma_\rho\gamma_\lambda = -4\varepsilon_{\mu\nu\rho\lambda}$$

and so

$$A(x) = Tr\gamma_5\left(\frac{1}{4}[\gamma_\mu, \gamma_\nu]F_{\mu\nu}\right)^2 \frac{1}{2}\int \frac{d^4k}{(2\pi)^4}\exp(-k^2) \qquad (4.7.15)$$

Put $d^4k = \pi^2k^2dk^2$ and we have

$$(1/2\pi)^4 \int d^4k \exp(-k^2) = 1/(16\pi^2). \qquad (4.7.16)$$

Then

$$A(x) = 1/(128\pi^2)Tr\gamma_5\gamma_\mu\gamma_\nu\gamma_\rho\gamma_\lambda F^{\mu\nu}F^{\rho\lambda}$$

$$= -1/(32\pi^2)\varepsilon_{\mu\nu\rho\lambda}F^{\mu\nu}F^{\rho\lambda} = -1/(16\pi^2)^*F_{\rho\lambda}F^{\rho\lambda} \qquad (4.7.17)$$

This shows that the measure

$$d\mu \to d\mu \exp[(i/8\pi^2)\int {}^*F_{\rho\lambda}F^{\rho\lambda}dx\alpha(x)]. \qquad (4.7.18)$$

From here some algebra leads us to Eq. (3.9b) for the anomaly.

We shall see in the next chapter that there are certain gauge field configurations in Euclidean space-time, called Instanton configurations, for which the classical action has local minima – and for these $^*F = F$. These configurations give a non-perturbative contribution to anomalous divergence.

5 Introduction to Instantons

5.1 Instantons or Pseudoparticles

In the quasi-classical approximation, i.e. for small coupling constants, the complicated vacuum structure of QCD can be shown explicitly. Along with the trivial vacuum sector corresponding to the vanishing of vacuum fields $A^a_\mu = 0$[1] (small oscillations near $A^a_\mu = 0$ are accounted for by perturbation theory) there are infinitely many other sectors in which the vacuum field $\left(\hat{A}^a_\mu\right)_{\text{vac}}$ yields $G^a_{\mu\nu} = 0$ and, still, cannot be reduced to $A^a_\mu = 0$ by any continuous gauge transformations. These additional sectors are labeled by integer numbers, the so-called winding numbers or topological charges. The corresponding classification of the non-equivalent vacuum sectors was first given by *Belavin, Polyakov, Schwarz* and *Tyupkin*, 1975, where the tunneling transition connecting the neighbouring sectors was also found. The field configurations interpolating between the classical vacua with different winding numbers is localized in space and in imaginary time.

A localized solution of Euclidean field equations with finite action is called a *pseudoparticle* in the original paper. Later the name *instanton* became more popular. The instanton contribution in the action is equal to $8\pi^2/g^2$, the corresponding amplitude is proportional to $\exp(-8\pi^2/g^2)$. This factor is evidently beyond perturbation theory. Even though instanton calculus has problems, it is useful. This is because at the moment it is practically the only concrete example of non-perturbative QCD fluctuations which, in a sense, serves as a probe element for different theoretical constructions.

To understand fluctuation between multiple vacua, take the simplest case of *two minima*: a double well potential. Consider two similar potential wells in the upper left and right quadrant of the x-energy plane. Let the solutions be

$$H_0|L\rangle = E_0|L\rangle \tag{5.1.1}$$

$$H_0|R\rangle = E_0|R\rangle \tag{5.1.2}$$

The H_0 is invariant under parity transform $x \Rightarrow -x$. But the states are not eigenvalues of the parity operation:

$$P|L\rangle = |R\rangle \tag{5.1.3}$$

$$P|R\rangle = |L\rangle \tag{5.1.4}$$

---

[1] Here the label a stands for the colour degree of freedom.

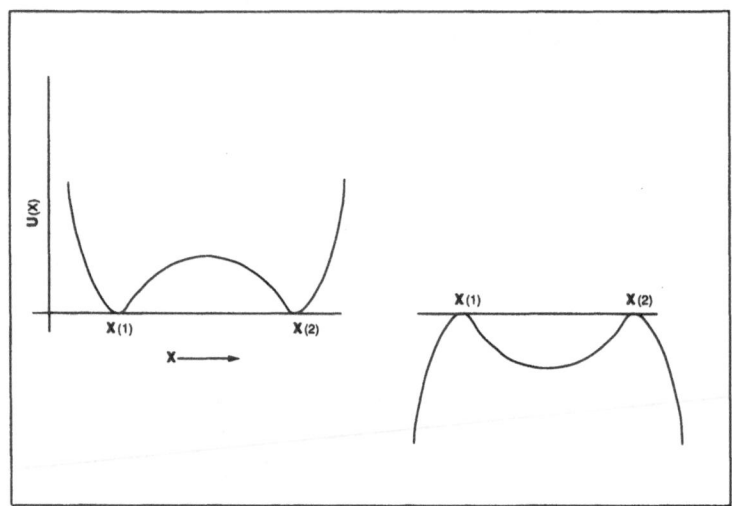

Fig. 5.1. Symmetric double well potential which changes sign as $t = x_0 \Rightarrow -ix_4$. Classically, a particle which was sitting either at $x(1)$ or at $x(2)$ can have solution, in the Euclidean space, which allows motion from $x(1)$ to $x(2)$

Let us now lower the barrier between the wells (*Bhaduri*, 1988) so that $H = H_0 + h$ and now tunneling is possible mixing the two states. Let

$$\langle R|h|L \rangle = \langle L|h|R \rangle = \delta \tag{5.1.5}$$

When one solves the eigenvalue equation one gets $E_0 \mp |\delta|$ as the energies of the states $|\psi_{S,A}\rangle = (1/\sqrt{2})\,(|L\rangle \pm |R\rangle)$ respectively. The symmetric state is obviously the ground state since by adjusting the height h one can make the barrier vanish, and then the symmetric state survives.

The question is: how does one know that there is the possibility of tunneling, using classical physics? The answer is to look at the Euclidean version $t = x_0 \Rightarrow -ix_4$. The well turns upside down. Recall that the energy is the fourth component of momentum. The finite action Euclidean solution is the one where the particle rolls from one hill to another. The potentials and this Euclidean version are illustrated in Fig. 5.1. We will discuss this in detail in the next section.

5.2 Tunneling in Imaginary Time

Consider a unit mass particle in a double well potential in real time. We can write the Lagrangian as $L = 1/2(\partial_0 q)^2 - 1/2 s^2(q) = 1/2(\partial_0 q)^2 - U(q)$. It will be very useful to recall *the kink* of Chapter (3.6) in what follows. There we had $U(\varphi) = 1/2 s^2(\varphi) = \lambda/4(\varphi^2 - \mu^2/\lambda)^2$ and $E \geq 0$ had a bound. Here instead of the field φ as the generalized coordinate we simply have q and Newtonian dynamics. Classically of course the particle will sit in one or other of the wells marked $L(R)$

with $q = \pm\mu/\sqrt{\lambda}$. The quantum tunneling can be considered *classically* if one goes to the action in the Euclidean version.

The corresponding Euclidean action is obtained by continuing $t = x_0 \Rightarrow -ix_4$:

$$S_{\text{Mink}} = \int L dt = -i \int dx_4 L(t \Rightarrow -ix_4)$$

$$= \frac{i}{2} \int dx_4 \left[\left(\frac{dq}{dx_4} \right)^2 + s^2(q) \right] \tag{5.2.1}$$

so that

$$S_{\text{Eucl}} = -i S_{\text{Mink}} = \int dx_4 L = \frac{1}{2} \int dx_4 \left[\left(\frac{dq}{dx_4} \right)^2 + s^2(q) \right] \tag{5.2.2}$$

Observe that the potential has changed sign (Fig. 5.1). By extremizing the action we get the Euclidean equation motion:

$$\frac{d^2 q}{dx_4^2} = \frac{1}{2} \frac{d}{dq} [s^2(q)] \tag{5.2.3}$$

Equation (2) can be written as

$$S_{\text{Eucl}} = \frac{1}{2} \int dx_4 \left[\left(\frac{dq}{dx_4} \pm s(q) \right)^2 \mp 2s(q) \frac{dq}{dx_4} \right]. \tag{5.2.4}$$

Hence the bound

$$S_{\text{Eucl}} \geq \int dx_4 | \frac{dq}{dx_4} s(q) | \tag{5.2.5}$$

When the bound is saturated we get the equation of motion (Eq. (3)). In other words, the classical trajectory saturates the bound.

Analogous to the <u>kink</u> solution of Chapter 3.6 we have finite action instanton solution $q(x_4) = \pm\sqrt{(\mu^2/\lambda)} \tanh(\mu x_4/\sqrt{2})$ with the action $S_{\text{Eucl}} = 4\mu^3/(3\lambda\sqrt{2})$. Like the kink this has a topologically conserved index associated with it. This solution is related to the quantum amplitude for the particle to tunnel from potential minimum $q = -\mu/\sqrt{\lambda}$ to the other minimum $q = +\mu/\sqrt{\lambda}$. The important point is that S_{Eucl} (Eq. (2)) is the same as the field energy of the $(1+1)$-dimensional field theory in the static case (Eq. (3.6.8)):

$$E = \int \mathcal{H} dx = \int dx \frac{1}{2} \left[\left(\frac{d\varphi}{dx} \right)^2 + s^2(\varphi) \right] \tag{5.2.6}$$

with Euclidean time x_4 and particle co-ordinate q replaced by x and the field φ. To be clear we state it again: Euclidean action for $(1+0)$ dimension is the same as the static solution for $(1+1)$-dimensional field theory. Instanton is a classical solution with finite action in Euclidean space. Solitons are finite energy solutions in classical field theory. So the instanton solution in $(d+1)$ dimension is the same as kink solution in $(d+1)$ space $+ 1$ real time.

5.3 Homotopy Theory

Mathematicians have found out the cases when there can be topological objects in a theory. This depends on the dimensions of the theory and on the group which describes the internal symmetries. We give a brief review of what we need, for more details see *Actor*, 1979.

We start with an example, a vortex in QED in (2+1) dimension.

$$\mathcal{L} = -\frac{1}{4}F_{\mu\nu}F_{\mu\nu} + \frac{1}{2}(D_\mu\varphi)^*(D_\mu\varphi) - \frac{\lambda}{4}(\varphi^*\varphi - a^2)^2 \tag{5.3.1}$$

here φ is a complex scalar field, $D_\mu = \partial_\mu - ieA_\mu$ and $F_{\mu\nu} = \partial_\mu A_\nu - \partial_\mu A_\nu$, the familiar electromagnetic tensor. The name vortex solution comes from the theory of superconductors. The Meissner effect consists of the fact that if an external magnetic field has strength smaller than a certain critical value H_0, then the field cannot penetrate inside the superconductor; if $H > H_0$ the field can go through a kind of hole through a superconductor of type II and there is a magnetic flux across the superconductor. These are called vortices of magnetic field which are quantized.

The Hamiltonian corresponding to Eq. (1) in the static limit with $A_0 = 0$, takes a form analogous to the free energy in Landau–Ginzburg theory of superconductivity. The latter is given by

$$G(\varphi, A_\mu) = \int dx \left[\frac{1}{2}|\nabla\varphi - ie\underline{A}\varphi|^2 + \frac{1}{4}\left[|\varphi|^2 - F(T)^2\right] + \underline{B}\cdot\underline{B}/(8\pi)\right],$$

$$\tag{5.3.2}$$

where $a = F(T)$. This was observed by *Nielsen* and *Olesen* (1973). In this, T is below the critical temperature T_c. We need a minimum two space dimension to define the magnetic field $\underline{B}$ (which is perpendicular to the space dimensions). Non-trivial topology plays a very important role here. Actually one gets flux quantization (see next section).

Zero energy configuration must have $A_i = e^{i\alpha(x)}$ for all x and $\varphi^*\varphi = a^2$. This condition leads $F_{ij}(x) = 0$ for all x. Hence the energy $E = 0$. For finite energy solutions, the energy density must be localized and therefore it is sufficient to have the condition $F_{\mu\nu} \Rightarrow \infty$ asymptotically. In other words as $|x| \Rightarrow \infty$, $\varphi^*\varphi \Rightarrow a^2$, so that at spatial infinity, $\varphi \Rightarrow ae^{i\alpha(x)}$. Since $x = \rho e^{i\theta}$, $|x| \Rightarrow \infty$, $\rho \Rightarrow \infty$. But θ still varies from zero to 2π. Thus spatial infinity maps to a circle and here α depends on θ only and asymptotically,

$$\varphi \Rightarrow ae^{i\alpha(\theta)} \tag{5.3.3}$$

To ensure $(D_i\varphi)^*(D_i\varphi) \Rightarrow 0$,

$$A_i = 1/e\partial_i[\alpha(\theta)] \tag{5.3.4}$$

so that $D_i\varphi = \partial_i\varphi - ieA_i\varphi = 0$ and $F_{\mu\nu} \Rightarrow 0$. Since the physical φ should be single-valued after a 2π rotation of θ:

$$\alpha(2\pi) - \alpha(0) = 2n\pi = \int_0^{2\pi}\frac{d\alpha}{d\theta}d\theta \tag{5.3.5}$$

Obviously $n = 0$ corresponds to a constant A_i. We get back $E = 0$, the trivial solution. For non-trivial n, $E > 0$. n must belong to different topological sectors and be a conserved quantity. Otherwise the solution will decay into $E = 0$ vacuum.

The transformation given by Eq. (3) is the element of the $U(1)$ group. For the complex field $\varphi = \varphi_1 + i\varphi_2, \varphi^*\varphi = \varphi_1^2 + \varphi_2^2 = 1$. The components φ_1 and φ_2 lie on a circle in the 2-d internal field space. Thus realization of $U(1)$ in the field space is a circle, S^1. For three scalar fields subject to $\Sigma\varphi_i^2 = 1$, we would get a spherical surface S^2. This is the case in non-linear $0(3)$ model as applied to isotropic ferromagnet (*Rajaraman*, 1982).

Let us consider the parameters of $SU(2)$. The group element is

$$U = e^{i\tau\alpha} = \cos\alpha + i\hat{\tau} \cdot \hat{\alpha}\sin\alpha = \sigma + i\hat{\tau} \cdot \underline{\pi}. \qquad (5.3.6)$$

We have

$$\text{Det}\, U = 1, U^\dagger U = 1, \sigma^2 + \underline{\pi}^2 = 1 \qquad (5.3.7)$$

σ and $\underline{\pi}$ lie on the hypersurface S^3. Any point on this corresponds to an element of $SU(2)$. Hence it is a manifold of $SU(2)$. σ and $\underline{\pi}$ should not be confused with the physical fields φ_i which are isoscalar and isovectors under $SU(2)$ transformation[2]:

$$\varphi_i \Rightarrow \varphi_i' = U\varphi_i \qquad (5.3.8)$$

5.4 Compactification of Space-Time Manifold and Mapping

The asymptotic boundary condition compactifies the field solutions at ∞. For example, boundary condition given by Eq. (3.3) maps 2-dimensional physical space $\mathbb{R}^2$ at spatial infinity into a circle S^1. We designate this by $S^1_{(phy)}$ to distinguish the $S^1_{(int)}$ of the circle in the internal field space. Solutions φ_i in $S^1_{(phy)}$ must correspond to a point on $S^1_{(int)}$. This is achieved by mapping the circles onto each other and is illustrated in Fig. 5.2. X is the topological space with a circle $S^1_{(phys)}$ and $S^1_{(int)}$ has to be mapped onto it.

The simplest possible mapping is a non - winding one: where all the points of $S^1_{(int)}$, including a point P, are mapped to some point N on $S^1_{(phy)}$; alternatively each point of $S^1_{(int)}$ is mapped to a different point on a loop l_0 as shown in figure. These mappings are homotopically equivalent because one can, by continuous deformation of l_0, get to the point N. The mapping l_1 in the figure is obviously distinct, one needs to cut it open to reduce it to l_0.

Mathematically, we take a manifold X with an element x, and a manifold Y with an element y. $f_0(x) = y$, $f_1(x) = y$ are homotopic if they can be deformed continuously into each other. The formal definition is as follows: if there exists

---

[2] In chiral perturbation theory or in Skyrmion, $U(x)$ itself is considered to be a field.

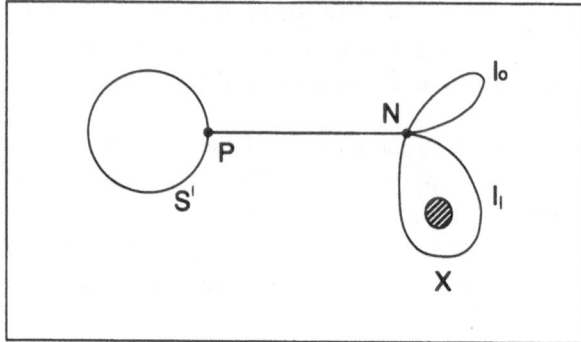

Fig. 5.2. S^1 can map the topological space (X) containing
the shaded region at a point N
or in a loop l_0. They are equivalent because none of them
covers the shade and a continuous deformation will transform
one to the other. But the loop
l_1 is distinct from them as it
winds the shade once

$F(x, t)$, $0 \le t \le 1$ such that

$$F(x, t = 0) = f_0(x), \quad F(x, t = 1) = f_1(x) \tag{5.4.1}$$

then the two mappings are homotopic to each other.
  Examples:

X	Y
θ	$\Lambda(\theta)$

Mapping:

1. $\Lambda_1(\theta) = 0$ for all θ.
2. $\Lambda_2(\theta) = t\theta$ for $0 \le \theta \le \pi$.
   $= t(2\pi - \theta)$ for $\pi \le \theta \le 2\pi$. (5.4.2)
3. $\Lambda_3(\theta) = \theta$ different.
4. $\Lambda_4(\theta) = n\theta$ different now for each n.

Here Λ_1 and Λ_2 are homotopic with winding number, $n = 0$. If the loop l_1
of Fig. 5.2 encircles the $S^1_{(phy)}$ once then $n = \pm 1$ depending on the winding
being anticlockwise or clockwise. Λ_3 above corresponds to $n = \pm 1$. Thus all
topologically equivalent mappings form an element of the homotopy group $\mathbf{Z}$ under
addition, with $n = 0$ as the identity element. Here $\Pi_1(S^1) = \mathbf{Z}$, subscript in Π
corresponds to $S^1_{(phy)}$.
  Consider Eq. (3.1) in (3 + 1) dimension in the static limit. The physical space
is now $\mathbb{R}^3$ but the internal symmetry at the spatial boundary is still $U(1)$. Thus the
boundary condition (Eq. (3.3)) compactifies $\mathbb{R}^3$ into S^2. Thus it is a mapping of
$S^2_{(phy)}$ onto $S^1_{(int)}$ forming a homotopy group $\Pi_2(S^1)$. For a finite energy topologically distinct solution we need the mapping to be non-trivial. We know that all the
circles on the surface of a sphere are homotopically equivalent and $\Pi_2(S^1) = 0$.
  Consider the case when the components of internal field space φ_i, ($i = 1, 2, 3$)
are constrained to lie on a spherical surface $S^2_{(int)}$, and the physical space is $\mathbb{R}^2$. φ
goes to a unit vector in internal space from all directions at the spatial boundary:
as $|\underline{x}| \Rightarrow \infty$ asymptotically $\varphi(\underline{x}) \Rightarrow \varphi^0$. This is compactification of $\mathbb{R}^2$ into S^2.

The physical space is folded into an $S^2_{(phy)}$, the circle at spatial boundary meets at a point. The mapping here from $S^2_{(int)}$ onto $S^2_{(phy)}$ is non-trivial. The gauge transformation has a dual role. It is known to lead to a unique theory with gauge invariance leading to renormalizability. It also helps us to construct topological objects using mapping from space-time to group manifold. We generalize from the above examples as follows:

Recipe: Gauge group G, $(n + 1)$-dimensional manifold $\Rightarrow S^n$ from asymptotics.
  If $\Pi_n(G) \neq 0$ then there are n-th homotopy group S^n to $S(G)$.
  Topologically non-trivial solutions are:
  $\Pi_n(S^1) \neq 0$ only for $n = 1$.
  $\Pi_1(S^1) = \mathbf{Z}$, the natural numbers.
  $\Pi_n(S^n) = \mathbf{Z}$
  $\Pi_n(S^m) = 0 \quad n < m$
  $\Pi_3(SU(2)) = \Pi_3(S^3) = \mathbf{Z}$

This is all we *need* to know. Let us go back to scalar electro-dynamics in (2 + 1) dimensions (Eq. (3.2)). We shall see that there *are* topological quantities leading to flux quantization. It is given by

$$\Phi = \int F_{12} \mathrm{d}^2 x \tag{5.4.3}$$

where F_{12} is the only component of the magnetic field $B_z = F_{12} = \partial_x A_y - \partial_y A_x$. Using Stoke's theorem we can convert this to a line integral:

$$\Phi = \oint \underline{dl} \cdot \underline{A} \tag{5.4.4}$$

As $\rho \Rightarrow \infty$, the tangential component of $\underline{A}$ (see Eq. (3.4)) becomes

$$A_\theta = \frac{1}{e\rho} \frac{\mathrm{d}\alpha}{\mathrm{d}\theta} \tag{5.4.5}$$

so that, using Eq. (3.5), the flux is

$$\Phi = \int A_\theta \rho \mathrm{d}\theta = 1/e[\alpha(2\pi) - \alpha(0)] = 2n\pi/e \tag{5.4.6}$$

The integer n is the winding number which describes the number of times the phase angle is wound around the circle at spatial infinity. It corresponds to $\Pi_1(S^1) = \mathbf{Z}$, $n \in \mathbf{Z}$.

Consider the conserved topological current $J_\mu = \varepsilon_{\mu\nu\lambda} \partial^\nu A^\lambda$, $J_0 = \varepsilon_{0ij} \partial^i A^j \Rightarrow F_{12}$. Thus $Q = \int J_0 \mathrm{d}^2 x = \int \varepsilon_{ij} \partial^i A^j \mathrm{d}^2 x \Rightarrow \Phi$ apart from an overall constant.

Application to Condensed Matter for the Interested Reader

1. Quantum Hall Effect $\Rightarrow$ anomaly in Schwinger model. [See *Semenoff* and *Sodano* (1986)].

2. Gapless semi-conductor gives exceptionally strong magnetic conductivity. See: The ABJ anomaly and Weyl fermions in a crystal, a long 8-page letter by *Nielsen* and *Ninomiya* (1983).
3. Chiral anomaly $\Rightarrow$ incommensurability in charge density (quasi-one dimension) wave to sinusoidal nonlinear coherent response, interpreted as a Josephson-type quantum oscillation is derived intuitively as an effect of explicit symmetry breaking in *Su* and *Sakita* (1986).

5.5 Bounds for the Instanton Solutions

Let us recall the bound in case of particle dynamics Eq. (2.5): $S_{\text{Eucl}} \geq \int dx_4 |\partial_4 q s(q)|$. When the bound is saturated one gets the equation of motion. For non-Abelian Euclidean gauge theory similar expressions follow (for details see for example, *Dittrich* and *Reuter*, 1985). Lagrangian and Hamiltonian densities of the gauge field in Minkowski space are:

$$\mathcal{L} = -\frac{1}{4}F^i_{\mu\nu}F^{i\mu\nu} = -\frac{1}{2}Tr F_{\mu\nu}F^{\mu\nu} = (E^2 - B^2)/2 \tag{5.5.1}$$

$$\mathcal{H} = (B^2 + E^2)/2 \tag{5.5.2}$$

where $F^{\mu\nu}$, $\underline{E}$ and $\underline{B}$ are matrix-valued functions. For example, $F^a_{\mu\nu}$ as in Eq. (2.1.4) can be written as

$$F^a_{\mu\nu} = \partial_\mu A^a_\nu - \partial_\nu A^a_\mu - ig[A_\mu, A_\nu]^a \tag{5.5.3}$$

Defining $F_{\mu\nu} = \Sigma_a F^a_{\mu\nu}T^a$ and $A_\mu = \Sigma_a A^a_\mu T^a$ we have the matrix-valued field tensor[3]

$$F_{\mu\nu} = \partial_\mu A_\nu - \partial_\nu A_\mu - ig[A_\mu, A_\nu] \tag{5.5.4}$$

Since covariant derivative $D_\mu = \partial_\mu - igA_\mu$, $F_{\mu\nu}$ can be expressed as the commutator of two D_μ:

$$[D_\mu, D_\nu] = -ig F_{\mu\nu} \tag{5.5.5}$$

The above equation leads to Bianchi identity:

$$D_\mu F_{\rho\sigma} + D_\rho F_{\sigma\mu} + D_\sigma F_{\mu\rho} = 0 \tag{5.5.6}$$

so, the dual field strength tensor $^*F_{\rho\lambda} = \frac{1}{2}\varepsilon_{\mu\nu\rho\lambda}F^{\mu\nu}$ satisfies

$$D^{\mu *}F_{\mu\lambda} = 0 \tag{5.5.7}$$

The equation of motion is given by

$$D^\mu F_{\mu\lambda} = 0 \tag{5.5.8}$$

---

[3] Notice that $-ig$ is not absorbed in the definition of A_μ. Hence the difference between Eq. (1) and Eq. (4.7.1.)

When $F_{\mu\lambda} = \pm\,^*F_{\mu\lambda}$, the equation of motion is always satisfied. Let us examine Eq. (1).

$$Tr F_{\mu\nu} F^{\mu\nu} = F^i_{\mu\nu} F^{j\mu\nu} tr(T^i T^j) = \frac{1}{2} F^i_{\mu\nu} F^{j\mu\nu} \delta^{ij} \qquad (5.5.9)$$

where $T^i s$ are the three generators, in case the group is $SU(2)$. For Minkowski to Euclidean, the replacements are: $x_0 \Rightarrow -ix_4$, $\partial_0 \Rightarrow i\partial_4$, $D_0 \Rightarrow iD_4$, $A_0 \Rightarrow iA_4$, $\underline{E}^{\text{Mink}} = i\underline{E}^{\text{Eucl}}$ and $\underline{B}^{\text{Mink}} = \underline{B}^{\text{Eucl}}$. Thus $\mathcal{L}$ and $\mathcal{H}$ get reversed when we go to Euclidean action:

$$S_{\text{Eucl}} = -iS_{\text{Mink}} = \frac{1}{4} \int d^4x F^i_{\mu\nu} F^i_{\mu\nu} = \frac{1}{2} \int d^4x Tr F_{\mu\nu} F_{\mu\nu}$$

$$= \frac{1}{2} \int d^4x (B^2 + E^2) \qquad (5.5.10)$$

Since $(B^2 + E^2) = (\underline{E} \pm \underline{B})^2 \mp 2\underline{E} \cdot \underline{B}$ we get the bound

$$S_{\text{Eucl}} \geq \int d^4x \underline{E} \cdot \underline{B} \qquad (5.5.11)$$

The bound is saturated when $\underline{E} = \pm\underline{B}$, or in other words,

$$F_{\mu\nu} = \pm\,^*F_{\mu\nu} \qquad (5.5.12)$$

One can check that the above solution gives zero Euclidean energy momentum tensor.

The pseudoscalar density $D = \frac{1}{4}F^i_{\mu\nu}\,^*F^i_{\mu\nu} = -\underline{E} \cdot \underline{B}$ can be expressed as a pure divergence:

$$D = \frac{1}{2g^2}\partial_\mu J_\mu \qquad (5.5.13)$$

where

$$\frac{1}{2g^2} J_\mu = \varepsilon_{\mu\alpha\beta\gamma} Tr \left(A_\alpha \partial_\beta A_\gamma + \frac{2g}{3i} A_\alpha A_\beta A_\gamma \right) \qquad (5.5.14)$$

Integrating over D, we get the topological charge or Pontryagin index, of the Euclidean field configuration A^a_μ:

$$q[A] = \frac{g^2}{8\pi^2} \int d^4x D(x) = \frac{g^2}{8\pi^2} \int d^4x \left(\frac{1}{4} F^i_{\mu\nu}\,^*F^i_{\mu\nu} \right) \qquad (5.5.15)$$

Note that q is gauge invariant. Using Eqs. (11) and (12) we obtain

$$S = \frac{8\pi^2}{g^2}|q| \qquad (5.5.16)$$

For finite S, $F_{\mu\nu}$ must go to zero at the boundary. This means for $|x| \Rightarrow \infty$

$$A_\mu \Rightarrow -\frac{1}{g}(\partial_\mu\Omega)\Omega^{-1} \qquad (5.5.17)$$

This boundary condition maps the 4-d space into S^3 and each point on S^3 associates an $SU(2)$ group element $\Omega(x)$, the manifold of which is again an S^3 (see Eqs. (3.6) and (3.7)). It defines a mapping

$$S^3 \Rightarrow SU(2), x \Rightarrow \Omega(x) \tag{5.5.18}$$

The solution satisfying Eq. (18) corresponds to $\Pi_3(S^3) = \mathbf{Z}$. And

$$q[A_\mu^a] = n \in \mathbf{Z} \tag{5.5.19}$$

We now consider the simplest non-trivial case, $n = 1$, the instanton. Instantons are localized, non-singular solutions of the classical Euclidean field equations, with one unit of topological charge. They are self dual and have a vanishing Euclidean energy momentum tensor. The associated action is

$$S[\text{instanton}] = \frac{8\pi^2}{g^2} \tag{5.5.20}$$

The instanton solution should also be regular at $r = 0$. The ansatz for $q = 1$, $SU(2)$ solution is:

$$A_\mu = \frac{r^2}{r^2 + \lambda^2}\left[-\frac{i}{g}(\partial_\mu\Omega)\Omega^{-1}\right] \xrightarrow[r^2 \Rightarrow \infty]{} -\frac{i}{g}(\partial_\mu\Omega)\Omega^{-1} \tag{5.5.21}$$

$$\Omega = \frac{\underline{x} \cdot \underline{\sigma} \mp ix_4}{|r|}, \quad \Omega^{-1} = \frac{\underline{x} \cdot \underline{\sigma} \pm ix_4}{|r|} \tag{5.5.22}$$

where $r^2 = x_4^2 + \underline{x}^2$ is the position and λ the size of the instanton. The generalization to colour $SU(N)$ was done by *Bernard*, 1979.

We mention that the importance of solutions like instantons, expressed in spaces with an imaginary time coordinate, is tied to the fact that only in such Euclidean space, integral forms essential to define generating functionals of Green's functions are well defined. In such theories the physical Green's functions result from the former by analytic continuation.

6 Relevance of Instantons

6.1 θ Vacuum

In real space-time $F_{\mu\nu} = 0$ (Eq. (5.1)) gives zero energy. This corresponds to a pure gauge field configuration for A_μ. Let us choose the temporal gauge $A_0 = 0$. Then the Minkowski-space vacuum solutions are time-independent pure gauge fields:

$$A_i(\underline{x}) = -\frac{i}{g}(\partial_i \Omega(\underline{x}))\Omega^{-1}(\underline{x}) \tag{6.1.1}$$

we restrict ourselves to a class of solutions for which

$$\Omega(\underline{x}) \Rightarrow 1 \text{ for } |(\underline{x})| \Rightarrow \infty \tag{6.1.2}$$

i.e. in all direction. This compactifies $\mathbb{R}^3$ to a three sphere S^3. This is a mapping of three space with all points at infinity to the identity element of the $SU(2)$ group of internal symmetry. Such mappings lead to the homotopy group $\Pi_3(SU(2)) = \mathbb{Z}$. Topological charge N characterizes different homotopy class. For example, $\Omega_0(\underline{x}) = 1$ yielding $A_i(\underline{x}) = 0$ belongs to $N = 0$ class. The $N = 1$ class corresponds to

$$A_i(\underline{x}) = e^{-\alpha^1(\underline{x})}\partial_i e^{\alpha^1(\underline{x})} \tag{6.1.3}$$

with

$$\alpha^1(\underline{x}) = \frac{i\pi\underline{\sigma} \cdot (\underline{x})}{(x^2 + a^2)} + i\pi\sigma_3 \tag{6.1.4}$$

The gauge transformation which takes the $N = 0$ configuration $A_i(\underline{x}) = 0$ to $N = 1$ configuration $A_i(\underline{x})$ is clearly

$$\Omega_1(\underline{x}) = e^{-\alpha^1(\underline{x})} \tag{6.1.5}$$

This will transform any classical vacuum in the N sector into one in the $(N + 1)$ sector. Or, in other words,

$$\Omega_N(\underline{x}) = [\Omega_1(\underline{x})]^N \tag{6.1.6}$$

We have therefore a distinct topological vacuum $|N\rangle$ in each sector. It is not the true vacuum since they will tunnel into each other as the system permits finite action instanton solution, $A_\mu(\tau, \underline{x})$, where $\tau = it$. The gauge invariant q (Eq. (5.5.17)) becomes

$$q = N_+ - N_- \tag{6.1.7}$$

where $N_\pm$ are topological charges for Minkowski vacua at $t = \pm\infty$ corresponding to Euclidean $\tau = \pm\infty$, i.e. Euclidean solution $A_\mu(\pm\infty, \underline{x})$ interpolates between Minkowski potentials in two different homotopy sectors, imaginary time τ being the interpolating parameter.

Since tunneling does take place, the physical vacuum – a linear superposition of topological vacua – will have $\theta \neq 0$:

$$|\theta\rangle = \sum_{-\infty}^{\infty} e^{iN\theta}|N\rangle, \qquad \theta \in [0, 2\pi] \tag{6.1.8}$$

$e^{iN\theta}$ makes the vacuum gauge invariant (upto a phase). This is because gauge transformation changes N. Suppose G is the corresponding operator to Ω_1 (Eq. (1.5)), then,

$$G|N\rangle = |N+1\rangle, \quad |N\rangle = G^N|0\rangle \tag{6.1.9}$$

$$G|\theta\rangle = \Sigma e^{iN\theta} G|N\rangle = e^{-i\theta}|\theta\rangle \tag{6.1.10}$$

It is easy to check that different θ vacua are orthogonal and no gauge invariant operator can generate transition between different θ vacua. If there exists such a vacuum the Lagrangian of a pure gauge field should contain an additional term (Eq. (5.5.10)):

$$i\theta \frac{g^2}{8\pi^2} \left(\frac{1}{4} F_{\mu\nu}^i {}^* F_{\mu\nu}^i \right) \tag{6.1.11}$$

Terms like above are odd under parity and even under charge conjugation and therefore CP-violating. So one could argue that we should set $\theta = 0$ in the strong sector. However this does not help since CP violating weak interactions renormalize θ to a non-zero value. How big is this θ_{QCD}? The answer is provided by a nuclear physics experiment.

6.2   The DEMON

The DEMON is the short name for the dipole electric moment of the neutron, d_n. Although chargeless, the neutron has a non-zero squared charge radius. But it should not have an electric dipole moment, since a non-zero d_n means both P and T violation. In nature CP non-invariance was discovered in the $K - \bar{K}$ system long time back. We refer readers to Lee's book (1981) for details. Different theories, while correctly describing K^0 decay, give widely varying predictions for d_n. As we have seen the Lagrangian of a QCD gauge field may have a term ($F \, {}^*F$) due to quantum fluctuation and this will lead to d_n. The DEMON was measured by *Altarev* et al. (1986) with the result:

$$d_n = (-1.2 \pm 0.4) \cdot 10^{-26} \text{ e cm.} \tag{6.2.1}$$

Another group (*K.F. Smith* et al. 1990) have confirmed that the order of this number is right. Although their result can only give limits and the source reactor of the experiment is shut down, it is this limit which is accepted by the particle data group. The value of d_n (Eq. (2.1)) calls for a θ_{QCD} of order 10^{-9}. There is a real challenge in making θ_{QCD} so small. One of the many approaches is the Peccei-Quinn mechanism (1977) where we make it actually zero. According to this, there is an approximate $U(1)$ global chiral symmetry realized with an almost massless pseudoscalar boson, the axion. Its mass m_a and the couplings

$$m_a = \frac{(m_q \Lambda_{QCD}^3)^{1/2}}{f_a}, \quad g_{aq\bar{q}} = \frac{m_a}{f_a} \tag{6.2.2}$$

are inversely scaled by a large axion decay constant f_a that is often related to large Higgs vacuum expectation value. Astrophysics, including most stringently the supernova 1987a, and cosmology tell us that

$$10^{11} \text{ GeV } \leq f_a \leq 10^{12} \text{ Gev.} \tag{6.2.3}$$

This leaves a narrow window in which the axion could provide an interesting and observable amount of dark matter. Detection of this dark matter would presumably rule out QCD as a source of large DEMON.

Let us refresh ourselves, for a change, with the description of one of the experiments, done at the VVR-M reactor at B.P. Konstantinov Leningrad Institute of Nuclear Physics in 1986. One keeps ultra-cold neutrons in prolonged confinement. The experiment was the so-called continuous-flow type. Two chambers, with oppositely directed electric fields and a magnetic-resonance spectrometer is used to find the differential count of flips. The neutrons are slowed in 1 litre of liquid hydrogen. Heating problem is severe. The neutrons generate 1200 Watts of heat removed by circulation. The hydrogen is placed at the core of the reactor.

We will discuss the possibilities whereby a DEMON is possible in electroweak theory or supersymmetric theories (SUSY in short). The standard electroweak model gives a d_n which is too small compared to that of Eq. (1). So it cannot be the only source for d_n.

(1) In SUSY every boson has an associated fermion. The gluon has the gluino. The gluino mass can have a phase, φ (see H.-P. Nilles, 1984, for a review). Normally when a fermion is introduced in any theory, its phase can be absorbed in the parameters of the Lagrangian. But the gluino is a Majorana spinor and for Majorana spinors the phase is non-trivial. This phase in turn may be related to $d_n = 10^{-22}\varphi$ (Buchmütller and Wyler, 1983, Polchinski and Wise, 1983). The φ is estimated to be of order α_s/π or about 0.05.

(2) The Weinberg multi-Higgs model. The two Higgs standard model leads to a d_n which is too small, so that Weinberg suggested a multi-Higgs model which was used by Khatsimovskii et al. (1987) to estimate the DEMON. They find a result which is too large, about 10^{-24}, so it appears the multi-Higgs model is also ruled out by present experiments.

(3) There are left-right symmetric SUSY models which predict results consistent with experiment, in that the DEMON is expected to be between 10^{-25} to 10^{-27} e cm.

On the whole the QCD θ parameter or the SUSY models seem to be the most favourable candidates for explaining the DEMON. If the DEMON exists it makes us ask the pressing questions: Why is the θ so small? Where is the gluino? Why is left-right symmetry broken in low energy everyday life and is it really observable at high energy?

6.3   Electroweak Baryon Number Violation

In 1976 't Hooft had already indicated that there will be $SU(2)$ instantons in the electroweak theory and that they would violate CP, and baryon number, B (or lepton number L) need not be conserved in that theory. But the suppression factor $\exp(-8\pi^2/g_w^2)$ becomes very small $\simeq 10^{-78}$ since the weak coupling g_w is very small. Interest in these phenomena was aroused again after the work of *Kuzmin*, *Rubakov* and *Shaposhnikov* (1985), who argued that B and L violating processes are much more probable at high temperatures. They used classical particle-like but unstable solutions of electroweak theory, called sphalerons, found earlier by *Soni* (1980) and *Klinkhamer* and *Manton* (1984) that resemble the instanton.

Standard electroweak theory is not a simple $SU(2)$ gauge theory. It is a spontaneously broken $SU(2)\text{X } U(1)$ gauge theory with massive bosons and a massless photon in a mixed state. One needs to use a *constrained instanton* description (*Affleck*, 1981) which looks like an instanton near its centre but decays exponentially with the corresponding mass scale far away from the centre. *Ringwald* (1990) has used this constrained instantons and computed the fermionic Green's function. In an attempt to include the effects of many external particles on the instanton itself, *McLerran*, *Vainshtein* and *Voloshin* (1990) have considered the effect of a constant external source J to account for possible distortion in the shape of the instanton. The main effect is to shift the vacuum expectation value of a scalar field $v \Rightarrow v_J$. They showed that in the limit of large J, even the distorted instanton yields a cross-section which has the same form as that obtained by Ringwald, although their considerations were limited to scalar bosons only. We cannot dwell further on the subject but refer the interested reader to the hotly pursued literature.

6.4   Conformally Invariant Solutions of Jackiw, Nohl and Rebbi

Conformal invariance has been discussed in the literature for a long time. Since we will be very brief we refer the reader to *Itzykson* and *Zuber* (1985, p. 642) and *Pokorski* (1987, p. 161). There is a long cherished notion, among symmetry minded physicists, that at high energy masses can be ignored, and a new symmetry sets in: the symmetry of scale or dilatation invariance. In addition, it is found, that in such theories with *no scales*, angles are also preserved and one gets the 15-parameter group of conformal transformation:

$$\left.\begin{array}{l}\text{translations } x'_\mu = x_\mu - a_\mu \\[4pt] \text{scale transformations } x'_\mu = e^{-\varepsilon} x_\mu \\[4pt] \text{Lorentz transformations } x'_\mu = \Lambda_{\mu\nu} x_\nu \\[8pt] \text{special conformal transformations } x'_\mu = \dfrac{x_\mu - b_\mu x^2}{1 - 2b \cdot x + b^2 x^2}\end{array}\right\}$$

where ε is a real number giving the scale transform and a, b are some four-vectors. Conformally invariant field theories have attracted lot of attention recently, for a review see *Sen* (1991). Since QCD vacuum does not have a scale we expect the vacuum solution to be conformally invariant. This led *Jackiw, Nohl* and *Rebbi* (1977) (JNR in short) to look for such solutions of the equations for the instanton. We show later that such solutions are highly relevant in connections with Skyrmion. We describe their notation first. The matrix-valued functions, action and Pontryagin densities are:

$$A_\mu = A^a_\mu \frac{\sigma^a}{2i}, \quad F_{\mu\nu} = F^a_{\mu\nu} \frac{\sigma^a}{2i} = \partial_\mu A_\nu - \partial_\nu A_\mu + [A_\mu, A_\nu]$$

$$S = \frac{1}{2} Tr F_{\mu\nu} F_{\mu\nu} \text{ and } {}^*S = -\frac{1}{2} Tr F_{\mu\nu} {}^*F_{\mu\nu}$$

where the coupling constant g is suppressed for convenience. $\sigma^a/2$ are the $SU(2)$ generators. Pontryagin index is given by

$$q = \frac{1}{2} \int d^4x \, {}^*S$$

It is useful to define a set of antisymmetric matrices $\bar\sigma_{mn}$ such that

$$\bar\sigma_{ij} = \frac{1}{4i}[\sigma_i, \sigma_j], \quad \bar\sigma_{i4} = -\frac{1}{2}\sigma^i, \quad i = 1, 2, 3 \tag{6.4.1a}$$

In terms of 't Hooft η symbol $\bar\sigma_{\mu\nu} = \eta^i_{\mu\nu} \dfrac{\sigma^i}{2}$ where $i = 1, 2, 3$ and

$$\eta^i_{\mu\nu} = -\eta^i_{\nu\mu} = \begin{cases} \varepsilon^{i\mu\nu} & \text{for } \mu, \nu = 1, 2, 3 \\ -\delta^i_\mu & \text{for } \nu = 4 \end{cases} \tag{6.4.1b}$$

These matrices are anti-self-dual $\bar\sigma_{\mu\nu} = -{}^*\bar\sigma_{\mu\nu}$ and the ansatz for the gauge field is $A_\mu = i\bar\sigma_{\mu\nu} a_\nu$ where a_ν is a vector field given by the self-duality condition $F_{\mu\nu} = {}^*F_{\mu\nu}$ as follows:

$$f_{\mu\nu} \equiv \partial_\mu a_\nu - \partial_\nu a_\mu = -{}^*f_{\mu\nu} \tag{6.4.2}$$

$$\partial_\mu a_\nu + a_\mu a_\nu = 0 \tag{6.4.3}$$

Note that the three potentials A^a_μ are expressed in terms of a single potential a_μ. Equations 2 can be satisfied if a_μ is derived from a scalar superpotential ρ:

$$a_\mu = \partial_\mu \ln \rho \tag{6.4.4}$$

$$A_\mu = i\bar\sigma_{\mu\nu} \partial_\mu \ln \rho \tag{6.4.5}$$

Then in terms of $\Box\rho \equiv \partial_\sigma\partial_\sigma\rho$, Eq. (3) gives

$$\frac{1}{\rho}\Box\rho = 0 \tag{6.4.6}$$

and the action density, which is now the Pontryagin density, may be expressed in terms of ρ:

$$-S = -{}^*S = \frac{1}{2}\Box\Box\ln\rho \tag{6.4.7}$$

In order that S be integrable, ρ must never vanish. If ρ is not singular, the Eq. (6) gives $\Box\rho = 0$ which permits only the trivial solution $\rho = $ constant. This corresponds to $q = 0$ but when ρ is singular we get interesting and ultimately non-trivial solutions for the gauge fields. For example, consider $\rho(x) = 1/|x|^2$ at $x \neq 0$, in 4 dimension.

$$\frac{1}{\rho}\Box\rho = \frac{1}{\rho}\partial_\sigma\left(-\frac{2x_\sigma}{|x|^4}\right) = 0 \tag{6.4.8}$$

Even at $x = 0$

$$\frac{1}{\rho}\Box\rho = -4\pi^2|x|^2\delta^4(x) = 0 \tag{6.4.9}$$

Therefore, the singular solutions of the form $\rho(x) = \lambda^2/(x - y)^2$ are acceptable since S remains regular at $x = y$. Here y are the 4 position coordinates of the instanton. In general,

$$\rho(x) = 1 + \sum_{i=1}^{n}\frac{\lambda_i^2}{(x - y_i)^2} \tag{6.4.10}$$

where λ_i and y_i are real numbers. This is the 't Hooft solution. These solutions when inserted in the A_μ give n instanton solution with $q = n$. First term corresponds to the $q = 0$ sector. For details see *Rajaraman, 1985*.

It is possible to get a conformally invariant form for the instanton solution. JNR wrote down a conformally invariant form:

$$\rho(x) = \sum_{i=1}^{N}\frac{\lambda_i^2}{(x - y_i)^2} \tag{6.4.11}$$

where $N = n+1$ and for the special case $y_N \Rightarrow \infty$ and $\lambda_N \Rightarrow \infty$ with $y_N/\lambda_N \Rightarrow 1$ one recovers the solution (8).

Pontryagin index $q = N - 1$. Let us now examine how many free parameters an instanton can have. In Eq. (10) for $n = 1$ there are 5 parameters, the size λ_i and four components of y_i. This corresponds to dilatation or scale transformation and translation symmetry respectively. The size λ_i can be arbitrary as long as it is non-zero. This is because when $x_\mu \Rightarrow \Lambda x_\mu$, the gauge potential $A_\mu \Rightarrow \Lambda A_\mu$ also. Solutions of Eq. (11) contain 4 more relevant parameters, since a common rescaling of the λ-s does not affect the expression of the A_μ. JNR showed that the additional degrees of freedom are not gauge artifacts. They have physical significance. For example, in case of $N = 2$, the one instanton solution, additional constraints appear reducing the number of free parameters to 5. In case of 2 instanton solution it is 13. The free parameters in general are $8n - 3$ out of which $3n - 3$ describe the relative orientations of the instantons in group space. Qualitatively this could be understood in the following way.

To derive Eq. (8) we chose a particular solution for A_μ. When globally rotated in $SU(2)$ group space one gets three extra degrees of freedom in addition to the five (for $n = 1$), so that the total number comes to 8. For n instantons it is $8n$. The overall gauge orientation is unnecessary so that we get a total of $8n - 3$. *Atiyah* et al. (1978, ADHM in short) proved this rigorously using powerful matrix methods. *Atiyah* and *Ward* (1977) have shown that there are no more solutions. We will next move on to a very recent and exciting subject: how to get to Skyrmions from instantons.

6.5 Instanton to Skyrmion

(i) Skyrmions from Instantons: Zero Temperature

The Skyrme model is a non-renormalizable, non-linear field theory of pions which provides a good approximation to low energy hadron physics. We will describe it as a topological soliton along with the non-topological soliton model in a later chapter. But even there we have to be brief, since the literature on Skyrmion is vast. So we refer the reader right now to the detailed review article by *Zahed* and *Brown* (1986).

The Skyrme solution is

$$U(\underline{x}) = \sigma(\underline{x}) + i\hat{\tau} \cdot \underline{\pi}(\underline{x}) = \exp[if(r)\hat{x} \cdot \tau], \quad r = |\underline{x}| \tag{6.5.1}$$

where $f(r)$ changes from π to zero as r varies from zero to ∞. This is the hedgehog solution in spin-isospin space accepted quite generally and gives a "baryon" centred at the origin. *Atiyah* and *Manton* (1989, AM in short) related this object with our known field theoretic instanton, a hedgehog in spin-colour space, in a remarkable way by time-integration.

Let $A_\mu(x)$ be the gauge field vector in $\mathbb{R}^4$ with topological charge n. Denote the line $x = (\underline{x}, t)$, with Euclidean t variable and $\underline{x}$ fixed, the time-line through $\underline{x}$. Integrating over the time-lines we get formally:

$$U(\underline{x}) = T \exp \left[- \int_{-\infty}^{\infty} A_t(\underline{x}, t)\, dt \right] \tag{6.5.2}$$

where T denotes time ordering. More explicitly to compute $U(\underline{x})$ one must introduce the auxiliary quantity $\tilde{U}(\underline{x}, t)$ and solve the equation

$$\frac{\partial \tilde{U}}{\partial t} \tilde{U}^{-1} = -A_t \tag{6.5.3}$$

along a time-line with initial data $\tilde{U}(\underline{x}, -\infty) = 1$ and $U(\underline{x}) = \tilde{U}(\underline{x}, \infty)$.

The instanton field $A_\mu(x)$ is given by Eqs. (4.1) and (4.5) in terms of $\rho(x)$, a solution of the Eq. (4.6). The time component of A_μ is:

$$A_t = \frac{1}{2}\frac{\nabla\rho \cdot \tau}{\rho} \qquad (6.5.4)$$

$\rho(x)$ for the 't Hooft one-instanton solution (Eq. (4.10)) is written as:

$$\rho(x) = 1 + \frac{\lambda^2}{(\underline{x} - \underline{X})^2 + (t - T)^2} \qquad (6.5.5)$$

We set $\underline{X} = 0$ to get a Skyrmion at the origin in $\mathbb{R}^3$. T is irrelevant, so we set it to zero. The solution for the potential is

$$A_t = i\underline{x} \cdot \tau \left(\frac{1}{t^2 + r^2 + \lambda^2} - \frac{1}{t^2 + r^2} \right) \qquad (6.5.6)$$

On performing the time integration (Eq. (2)) we get the spherically symmetric hedgehog (Eq. (1)) with Baryon number $B = n$ as

$$f(r) = \pi \left(1 - \left(1 + \frac{\lambda^2}{r^2} \right)^{-1/2} \right) \qquad (6.5.7)$$

This has the right boundary condition for $n = 1$ and it is a good approximation to the numerical solution of actual Skyrmion for a suitable size λ. The energy for this is evaluated, using the Skyrme energy functional of *Adkins* et al. 1983, as a function of λ with *energy unit* $3\pi^2 F_\pi/\varepsilon$ and length unit $2/\varepsilon F_\pi$. The energy minimum occurs at $\lambda^2 = 2.11$ with energy 1.243 as opposed to the exact Skyrmion value 1.232. Here ε and F_π are related to the coefficients of the two terms in the Skyrme Lagrangian. Notice we need not even write down the Skyrme Lagrangian.

What is the reason behind this magical agreement? The reason may be the scale invariance of the instanton solutions, which gives the right boundary conditions and topology to a whole class of solutions (a one-parameter class, λ) and we are merely choosing the right one out of them to get close to the exact solution. However for this no *dynamics* is necessary, and therefore details of the solution are still likely to be different. For example, the profile $f(r)$ has the usual $1/r^3$ dependence for large r, but the coefficient is rather large.

To obtain a Skyrmion with another orientation let us use the JNR solution (4.11) for an instanton ($n = 1$, $N = 2$):

$$\rho(x) = \sum_{i=1}^{2} \frac{\lambda_i^2}{(x - X_i)^2} \quad \text{centred at } X = \frac{\lambda_2^2 X_1 + \lambda_1^2 X_2}{\lambda_2^2 + \lambda_1^2} \qquad (6.5.8)$$

with size $\lambda = \lambda_1\lambda_2/(\lambda_1^2 + \lambda_2^2)|X_2 - X_1|$.

Skyrmion, therefore, will be centred at $\underline{X}$ with this scale. Note that if $\lambda_1 \gg \lambda_2$, then the Skyrmion is centred near $\underline{X}_2$ and its scale size is much smaller than $|X_2 - X_1|$.

The last remarks suggest how to construct a configuration of two Skyrmions. Take the JNR solution for $n = 2$ instantons. We have three poles at X_1, X_2 and

X_3 with weights λ_1, λ_2, λ_3 respectively. Assume that the poles X_2 and X_3 are well separated in $\mathbb{R}^4$ with much smaller weights than weight at the pole X_1. The field is approximately that of the two single instantons centred at and concentrated near X_2 and X_3 (Eq. (8)). Corresponding $B = 2$ Skyrmion will be approximately the product of two single Skyrmions located at $\underline{X}_2$ and $\underline{X}_3$ in $\mathbb{R}$. Their scale sizes are much smaller than the distance between them, direction of orientation given by the line connecting X_2 to X_1 and X_3 to X_1. Thus one gets a "dibaryon" that looks like the product of two single Skyrmions. This was assumed in Skyrme's original paper and for example in *Jackson* et al. (1985). However, in general the fields are quite complicated and not the simple product, i.e. not two uncorrelated Skyrmions!

(ii) Skyrmions from Instantons: Finite Temperature

We shall introduce temperature and show that it fits in beautifully in the path integration formalism for quantum field theory. Quantum mechanically we find the transition amplitude of a particle from state $|x_1\rangle$ at $t = 0$ to a state $|x_2\rangle$ at t is given by Green's function $\langle x_2, t | x_1, 0 \rangle = \langle x_2 | e^{-iHt} | x_1 \rangle$. Going over to Euclidean time $\tau = it$ and using path integrals the amplitude is:

$$\langle x_2, \tau | x_1, 0 \rangle = \langle x_2 | e^{-H\tau} | x_1 \rangle = \int_{x_1(0)}^{x_2(\tau)} D[x] e^{-S} \tag{6.5.9a}$$

where D stands for the measure of the integral over all paths. If we take $|x_1\rangle = |x_2\rangle$ and sum, we shall have the trace. In field theory with $\mathcal{L} = \mathcal{L}(\varphi_i, \varphi_i^\mu)$ the vacuum to vacuum amplitude is written as (Eq. (4.4.2))

$$\langle 0_+ | 0_- \rangle = \int D[\varphi_1] D[\varphi_2] \ldots e^{-\int \mathcal{L}_E d^4 x} \tag{6.5.9b}$$

These ideas go over to statistical physics very easily. The partition function is a trace in Fock space $Z = Tr[\exp(-\beta H)]$ where the temperature $T \equiv 1/\beta$, controls the fluctuation instead of $\hbar$. For the same $\mathcal{L}$

$$Tr e^{-\beta H} = \int D[\varphi] e^{-\int \mathcal{L}_E d^4 x} \tag{6.5.9c}$$

with $0 \le \tau \le \beta$. Since the above is a trace, the initial and final states are the same and we must have the periodicity:

$$\varphi(\tau + \beta) = \varphi(\tau) \tag{6.5.10a}$$

Thus the statistical sum can be expressed as a sum over closed path in imaginary time. From the scalar fields given above one can obviously generalize to vector fields in a trivial way

$$A_\mu(\tau + \beta) = A_\mu(\tau) \tag{6.5.10b}$$

and to spinorial fields which is not so trivial (*McLerran*, 1986):

$$\psi(\tau + \beta) = -\psi(\tau) \tag{6.5.10c}$$

the functions now being anti-periodic. In momentum space one finds that the fourth component of the momentum becomes *discrete*:

$$\omega_k = i\omega_0 k \quad \text{(bosons)}$$

$$= i\omega_0 \left(k + \frac{1}{2} \right) \quad \text{(fermions)} \tag{6.5.11}$$

where k is an integer and $\omega_0 = 2\pi T$ is the so-called Matsubara frequency. The Fourier-type path integrals are changed to the sums according to the rule $(2\pi i)^{-1} \int d\omega \Rightarrow T\Sigma_k$. At $T = 0$ the Matsubara frequency disappears and we are returned to the ground state.

After this introduction to finite temperature we turn back to instantons, and generalize them to finite temperature as done by *Harrington* and *Shepard* 1978. Finite temperature instantons were called calorons by them. Recall one instanton solution of 't Hooft (Eq. (4.10)). Let us assume that the position coordinates of the instanton are $(0, 0, 0, \tau_0)$. Then the generalization at finite temperature $T = 1/\beta$, is simply:

$$\rho(x) = 1 + \sum_{k=-\infty}^{\infty} \frac{\lambda^2}{\underline{x}^2 + (\tau - \tau_k)^2} \tag{6.5.12}$$

where the poles are at $\tau_k = \tau_0 + k\beta$. Clearly ρ is periodic (Eq. (10)) in the "physical region" $0 \leq \tau - \tau_0 \leq \beta$. More generally, for n instanton one can replace β by β/n and write $\tau_k = \tau_0 + k\beta/n$. The above summation can be performed and after some algebra it is:

$$\rho(x) = 1 + \frac{\tilde{\lambda}}{4p} \left[\coth \frac{(p + i\tilde{\tau})}{2} + \coth \frac{(p - i\tilde{\tau})}{2} \right] \tag{6.5.13}$$

where $\tilde{\lambda} = 2\pi\lambda/\beta$, $p = 2\pi|\underline{x}|/\beta$ and $\tilde{\tau} = 2\pi(\tau - \tau_0)/\beta$.

That the solution will be given by coth function, can be seen very easily. Look for spherical solutions $\rho(r, \tau)$ like $\rho = f(r, \tau)/r$ so that

$$\Box\rho \equiv \ddot{\rho} + r^{-2} \frac{\partial}{\partial r} \left(r^2 \frac{\partial}{\partial r} \rho \right)$$

$$= (\ddot{f} + f'')/r = 0 \tag{6.5.14}$$

The general solution of (14) is thus of the form $\rho = [f_1(r + i\tau) + f_2(r - i\tau)]/r$. Requiring ρ to be real implies $f_1^*(z) = f_2(z^*)$. Combining this with the requirements that ρ be dimensionless, periodic, finite and positive we are led to a coth function.

Using the above solution in the analogue of Eq. (4.7):

$$S = -\frac{1}{2g^2} \int d^4x \Box\Box \ln\rho$$

we get for a spatial volume $V = (4\pi/3)R^3$

$$S_{R\Rightarrow\infty} = \frac{8\pi^2 n}{g^2} \left(1 - 10\pi^2 n^2 T^2 \lambda^4/R^2 \right) \tag{6.5.15}$$

where the validity of the expression depends on the assumptions that $R \gg \beta$ and that the second term in the parenthesis is small compared with 1. For more details the reader should read the original article which is short and clear.

Nowak and *Zahed*, 1989 extended the Atiyah-Manton procedure to obtain Skyrmion from the instanton at non-zero T. The application of Eq. (13) with the approximations made in (15) implies, that at finite T, $\lambda(T)$ goes to $\lambda[1 + (\pi \lambda T)^2/3]^{-1/2}$ leading to *shrinking* of the instanton size *and of the corresponding Skyrmion*. The fact that instantons shrink under the effect of T is not prohibited by the Yang-Mills equations since they are scale invariant. But the Skyrme Lagrangian is an effective one with field gradients. As the Skyrmion shrinks, field gradients become high, and should be suppressed.

At $T = 0$ the Skyrmion energy given by *dynamics* is $M = \varepsilon/\lambda + F_\pi^2 \lambda$, so that $(dM/d\lambda) = 0 \Rightarrow$ the equilibrium size $\lambda_m = \varepsilon^{1/2}/F_\pi$. One can invoke pion dynamics and finds the size increases with T, (see *Nowak* and *Zahed*, 1989; *Dey* et al. 1987 as well as *Gasser* and *Leutwyler*, 1987). One can of course argue that in any case the terms quartic in momentum of the pion field are numerous and divergent, as we shall see in the next chapter, so that there is no basis for Skyrme model from the derivative expansion. The only basis for Skyrme theory is phenomenology at $T = 0$ and the $T \neq 0$ extension should be viewed with caution.

The lesson is simple, one should be careful in using effective theories to predict new phenomena!

6.6   QCD Vacuum – Instanton Gas or Liquid?

Let us discuss some other instanton-induced effects in strong interaction physics.

If the instanton size was small it would imply strongly localized fluctuation. In this case one could place one region of fluctuation apart from another, assume no interaction between them and do a dilute gas calculation. But there is no *a priori* reason why the size of the instanton, λ, should be small. Hence there may be overlapping fluctuations and a free instanton will eventually be melted away. A certain short range repulsion between them may stabilize the instanton. Along this line *Shuryak* (1988a) proposed the instanton liquid model of the QCD vacuum. There are two parameters in the model, the instanton density, n_{pp} and the typical instanton radius λ_c. Obviously, $n_{pp} \neq 0$ will give rise to non-vanishing gluon condensates, $\langle 0| \left(g G^a_{\mu\nu} \right)^2 |0 \rangle$, where the field tensor $G^a_{\mu\nu}$ has non-trivial q (Eq. (5.5.16)). QCD sum rule analysis, (to be given in the last chapter) connecting different correlation functions to hadron phenomenology, extracts a value:

$$n = \frac{\langle 0| \left(g G^a_{\mu\nu} \right)^2 |0 \rangle}{32\pi^2} = (197 \text{ MeV})^4 \qquad (6.6.1)$$

Setting an upper limit for $n_{pp} < n$, *Shuryak* (1988) attempted to explain the non-perturbative effects in the spin-zero channel with a $\lambda_c \simeq 1/3$ fm. The instanton-instanton interaction was found out not to be negligible. Thus, if an instanton is stable at the scale λ_c, the QCD vacuum looks like a liquid rather than a dilute gas of non-interacting instanton.

6.7 Instanton Suppression by Light Fermions

The presence of massless fermions changes the picture. It was shown by *'t Hooft* (1976b) that the Dirac equation in an instanton background field has zero energy eigen value.

$$\mathbb{D}\Psi_0 \equiv (\partial_\mu + A_\mu)\gamma_\mu \Psi_0 = 0 \qquad (6.7.1)$$

where $\gamma_4 = i\gamma^0$ and the background $A_\mu(x)$ is the field configuration due to one instanton (Eq.(4.5)). For an instanton of size λ, located at the origin of $\mathbb{R}^4$, the zero mode solution Ψ_0 is given by (*'t Hooft* 1976b):

$$\Psi_0^a(x) = \frac{\lambda}{\pi(x^2 + \lambda^2)^{3/2}} \frac{\gamma \cdot x}{x} \begin{pmatrix} \phi^a \\ -\phi^a \end{pmatrix} \qquad (6.7.2a)$$

where a is the $SU(2)$ colour index. ϕ^a, the two component spinor is a hedgehog in colour-spin space:

$$(\sigma + \tau)\phi = 0 \qquad (6.7.2b)$$

where $\sigma(\tau)$ is the spin(colour) $SU(2)$ matrices. Ψ_0 has a definite chirality: $\gamma_5 \Psi_0^a(x) = \Psi_0^a(x)$. In general there could be n_+ number of zero eigenvalues of the operator $\mathbb{D}$ with positive and negative chirality. These $n_\pm$ numbers are related to the Pontryagin index q (Eq. (5.5.15)): $q = n_+ - n_-$. This is the Atiyah-Singer index theorem (1968). For $q \neq 0$, both n_+ and n_- cannot disappear. We shall see presently that it has a far reaching consequence on the instanton model of QCD vacuum. Recall that there will be tunneling between different topological vacua $|N\rangle$ (Section 1). This amplitude (Eqs. (5.9)) is given by:

$$\underset{\tau \to \infty}{\text{Lim}} \langle N + q|e^{-H\tau}|N\rangle = \int D[\Psi]D[\bar{\Psi}]D[A_\mu]e^{-S_E} \qquad (6.7.3a)$$

with $S_E = S_A + S_\Psi$, S_A is the action for the gauge fields. For a fermion of mass m the action

$$S_\Psi = i \int \bar{\Psi}(\mathbb{D} - m)\Psi d^4x \equiv \int \Psi^\dagger M[A_\mu]\Psi d^4x;$$

with $M[A_\mu] = i\gamma_4(\mathbb{D} - m)$ \qquad (6.7.3b)

Using Eq. (3b), the transition amplitude

$$\underset{\tau \to \infty}{\text{Lim}} \langle N + q|e^{-H\tau}|N\rangle \propto \int D[A_\mu]e^{-S_E} \text{Det}(M[A_\mu]) \qquad (6.7.3c)$$

$\text{Det}(M[A_\mu])$ comes in the measure itself [Eqs. (4.6.6–7)]. We see that the operator $M[A_\mu]$ has at least one zero eigen value for $m = 0$ limit, leading to $\text{Det}(M[A_\mu]) = 0$. Thereby the above tunneling probability gets suppressed in presence of nearly massless quarks.

Another important observation is that the instanton possesses the mode with one chirality while the anti-instanton, the other one. An instanton is accompanied with production of a quark pair of opposite chiralities, e.g. $\bar{q}_R q_L$. So, in a massless theory it is impossible to close them up in a loop. The strange quark has a more substantial mass and may be closed up.

What sort of picture emerges then for the QCD vacuum? In it there could be instanton clustering if only due to a kind of quark exchange interaction between them. The cluster may be like a molecule made up of an instanton (I) and an anti-instanton (A) as atoms and virtual quark pairs behaving like bonds living in 4 space dimension! Or, the cluster could be a polymer type as advocated by *Shuryak* (1988). The reason is: in a polymer of I and A the interaction generated from the IA between quarks *spontaneously break the chiral symmetry*, (SBCS in short). In a nut shell, SBCS means that even if the current quark mass $m \Rightarrow 0$, the vacuum expectation value (V.E.V., in short) $\langle \bar{\psi}\psi \rangle \neq 0$. As a result, the quark becomes massive. Attempts have been made by *Caldi* (1977), *Callan, Dashen* and *Gross* (1978), *Carlitz* and *Creamer* (1979) to see whether instantons are responsible for this. *Shuryak* (1988, c) and *Shuryak* and *Verbaarschot* (1991) showed that by fixing the instanton size to λ_c, $\langle \bar{\psi}\psi \rangle \neq 0$ when the structure of the vacuum is of polymer-type and $\langle \bar{\psi}\psi \rangle \Rightarrow 0$ as the structure changes from polymer to the molecular. Molecular vacuum is analogous to an "insulator" and the polymer-type vacuum to a "conductor" of axial current. Thus a new scale λ_c is defined in QCD vacuum above which the SBCS takes place. This scale is much smaller than the confinement scale, ~ 1 fm. For a value of $\langle \bar{\psi}\psi \rangle \cong -(255 \text{ MeV})^3$ (QCD sum rule result) the light quark attains an effective mass of ~ 300 MeV at a distance of $1/3$ fm., as

$$m_{\text{eff}} = m_q - \frac{2}{3}\pi^2 \lambda_c^2 \langle \psi \bar{\psi} \rangle \tag{6.7.4}$$

6.8   Instanton Induced Effective Interaction

It is now pertinent to ask how an instanton of small size develops in this physical vacuum which is an instanton liquid. With finite $\langle \bar{\psi}\psi \rangle$, vacuum – vacuum transition amplitude does not vanish. It can be expressed in terms of a local effective Lagrangian which consists of a $2N_f$ quark vertex (*'t Hooft*, 1976a):

$$\mathcal{L}_{\text{eff}} \propto \Pi \bar{\psi}(x) i\gamma \cdot \partial \psi^f_{(0)}(x) \bar{\psi}^f_{(0)}(x) i\gamma \cdot \partial \psi(x) \tag{6.8.1}$$

Note that only the zero mode solution has been kept as it dominates if the quarks are nearly massless. $\psi_{(0)}$ is given by the Eq. (7.1) and f labels each massless flavour. We require it to be a colour singlet and therefore colour averaging is necessary (see *Shifman* et al. 1980 for details). The spinor ϕ (Eq. (7.2)) describes a state for which the colour and usual spin add to each other to give a singlet so that

$\phi_{m\alpha} \sim \varepsilon_{m\alpha}$ $(m, \alpha = 1, 2)$. The density matrix for the state described by ϕ is of the form:

$$P_{mn,\alpha\delta} = \phi_{m\alpha}\phi_{n\delta}^\dagger = \frac{1}{4}(I_{mn}I_{\alpha\delta}^c - \sigma_{mn}\tau_{\alpha\delta}) \qquad (6.8.2)$$

where I and I^c are unit matrices in spin and colour space. In compact notation

$$P = \frac{1}{4}(I * I^c - \sigma * \tau) \qquad (6.8.3)$$

so that the V.E.V. of $\langle P \rangle_{SU(2)} = {}^1\!/4 I * I^c$ as for the colour vector τ, $\langle 0|\tau|0 \rangle = 0$. Next we have to average over all possible ways of embedding a particular $SU(2)$ into $SU(3)$. We have

$$(I^c)_{SU(2)} = \begin{pmatrix} 1 & 0 & 0 \\ 0 & 1 & 0 \\ 0 & 0 & 0 \end{pmatrix} = \frac{2}{3}(I^c)_{SU(3)} + \frac{\sqrt{3}}{2}\lambda^8 \qquad (6.8.4)$$

where λ^i are the 8 Gell-Mann $SU(3)$ matrices (given in Chapter 2). Hence $\langle P \rangle_{SU(2)} = {}^1\!/6 I * (I^c)_{SU(3)}$. Without going into further details we give the results of colour averaging (*Oka* and *Takeuchi*, 1991). For 2-flavour, $SU(2)$ colour quarks:

$$\langle \phi_1\phi_1^\dagger\phi_2\phi_2^\dagger \rangle_{SU(2)} = \left(\frac{1}{4}\right)^2 \left[1 + \frac{1}{3}(\sigma_1 \cdot \sigma_2)(\tau_1 \cdot \tau_2)\right] \qquad (6.8.5)$$

For $SU(3)$ colour and 2-flavour

$$\langle \phi_1\phi_1^\dagger\phi_2\phi_2^\dagger \rangle_{SU(3)} = \left(\frac{1}{6}\right)^2 \left[1 + \frac{3}{32}\lambda_1 \cdot \lambda_2 + \frac{9}{32}(\sigma_1 \cdot \sigma_2)(\lambda_1 \cdot \lambda_2)\right] \qquad (6.8.6)$$

For $SU(3)$ colour and 3-flavour

$$\langle \phi_1\phi_1^\dagger\phi_2\phi_2^\dagger\phi_3\phi_3^\dagger \rangle_{SU(3)} = \left(\frac{1}{6}\right)^3 \left[1 + \frac{3}{32}(\lambda_1 \cdot \lambda_2 + \text{perm.}) - \frac{9}{320}D_{123} + \right.$$

$$\left. \frac{9}{32}(\sigma_1 \cdot \sigma_2\lambda_1 \cdot \lambda_2 + \text{perm.}) + \text{three body terms}\right] \qquad (6.8.7)$$

In terms of $\psi_L = {}^1\!/2(1 - \gamma_5)\psi$ and $\psi_R = {}^1\!/2(1 + \gamma_5)\psi$, the left handed and right handed Dirac operators, the effective Lagrangian or the effective interaction, $H^{(3)}$ for the full $SU(3)$ colour and $SU(3)$ flavour becomes:

$$H^{(3)} = -\mathcal{L}_{\text{eff}} = V_0\bar{\psi}_R(1)\bar{\psi}_R(2)\bar{\psi}_R(3)\left[1 + \frac{3}{32}(\lambda_1 \cdot \lambda_2 + \text{perm.})\right.$$

$$- \frac{9}{320}D_{123} + \frac{9}{32}(\sigma_1 \cdot \sigma_2\lambda_1 \cdot \lambda_2 + \text{perm.})$$

$$\left. + \text{3-body terms}\right]\psi_L(3)\psi_L(2)\psi_L(1) \qquad (6.8.8)$$

where V_0 is the overall strength of the interaction. Assuming a quark condensate, we can extract the 2-body interaction:

$$H^{(2)} = -\mathcal{L}_{\text{eff}}^{(2)} = V_0^{(2)}\bar{\psi}_R(1)\bar{\psi}_R(2)\left[1 + \frac{3}{32}(\lambda_1 \cdot \lambda_2)\right.$$

$$+ \frac{9}{32}(\sigma_1 \cdot \sigma_2 \lambda_1 \cdot \lambda_2) \Big] \psi_L(2)\psi_L(1) + \text{h.c.} \qquad (6.8.9)$$

where $V_0^{(2)} = V_0 \langle \bar{\psi}_R(3)\psi_L(3) \rangle = \frac{1}{2}V_0 \langle \bar{\psi}\psi \rangle \propto m_{\text{eff}}^{(3)}$

Attempts have been made to examine the instanton induced interaction (III, in short) in hadron spectroscopy (*Horn* and *Yankielowicz* 1978; *Kochelev* 1985; *Shuryak* and *Rosner* 1989; *Dorokov* and *Kochelev* 1990a; *Dey* 1991; *Dey*, *Dey* and *Volkovitsky* 1991).

Why should we bother about such effects? Colour-spin factor in (9) is like that of one gluon exchange (OGE, in short) potential between the quarks. The central part of the OGE in the Fermi-Breit form:

$$V_{\text{OGE}} = \frac{\alpha_s}{4} \sum_{i<j} \frac{2\pi}{9m_i m_j} (\lambda_i \cdot \lambda_j)(\sigma_i \cdot \sigma_j)\delta(\underline{r}_{ij}) \qquad (6.8.10)$$

De Rujula, *Georgi* and *Glashow* (1975) and later *Isgur* and *Karl* (1978, 1979) used the above for the spin splitting of mesons and baryons. Even in MIT bag model $N - \Delta$ splitting is obtained from the color magnetic interaction of one gluon exchange. The main ingredient is that $\sigma_1 \cdot \sigma_2$ is repulsive for $S = 1$ states and attractive for $S = 0$ states. Sometimes one needs an α_s which is too large for perturbative treatment. Hence the need for the III which was derived from the non-perturbative multi-gluon configuration. In general both one gluon exchange and the III contribute towards the spin splitting thereby α_s required is smaller. We shall see that III contribute non-trivially towards hadron phenomenology.

(1) $\pi - K - \eta - \eta'$ mass difference was explained (*Dorokov* and *Kochelev*, 1990a). The bag model wave functions obtained with current quark mass are used to evaluate both the OGE and III. In fact, in the $m_u = m_d = m_s = 0$ limit, the splitting between pseudoscalar singlet and the octet arises exclusively from the III interaction (*Kochelev*, 1985). The effective interaction (Eq. (9)) is not invariant under $U(1)$ transformation. As a result, η' mass comes out to be high in a natural way. The symmetry breaking term is provided by the instanton.

Attempts to describe the spin-dependent proton structure functions via instanton contribution were made in *Dorokov* and *Kochelev* (1990b) and *Forte* (1989, 1990). A more pheneomenological approach was taken by *Ioffe* and *Karliner* (1990) who suggested helicity-dependent strange quark distribution in the proton with different large x behaviour obtained from possible non-perturbative contribution. Their conjecture can be tested (see *Preparata* et al. 1990). New experimental data from Fermilab Tevatron (*Rabinowitz* et al. preprint 1992) may decide the issue.

(2) In the frame work of QCD sum rule, $\pi - K - \eta$ masses, $\eta\eta'$ mixing as well the $U(1)$ breaking effect, η' mass, have been estimated using the instanton model of QCD vacuum (*Shuryak*, 1983). We shall discuss them in the last chapter.

(3) The full 3-body interaction (Eq. (8)) was used to find out the effect of III in the binding of H-dibaryon state (*Takeuchi* and *Oka*, 1991). The instanton effect plays a very important role. The V_{OGE} interaction with an α_s obtained by fitting the hyperfine splitting in baryon phenomenology overbinds the H-dibaryon. With the addition of the III it comes out to be barely bound. Including the meson exchange

the equations are solved using 6 oscillator shell and the resonating group method. However relativistic corrections are not considered and may be substantial.

(4) We apply the idea of instanton physics to look at the ground state properties of baryons. Recently by *Shuryak* and *Rosner* (1989) showed that the light baryon spectrum can be reproduced equally well, as by the V_{OGE} essentially because III is attractive for $S = 0$ states. Let us suppose that between a $(u\ d)$ pair it is α and between a $(u\ s)$ or $(d\ s)$ pair it is β. The ratio of α and β is not arbitrary. Look at the interaction strength $V_0^{(2)}$ (Eq. (9)). The third particle effective mass appears in the strength itself. Then with the masses of the quarks U and S (including all other interaction except the $S = 0$ attraction due to III) one can write:

$$\alpha/\beta = S/U \tag{6.8.11}$$

and

$$N = 3U - 3\alpha/2. \tag{6.8.12a}$$

$$\Lambda = 2U + S - \alpha - \beta/2. \tag{6.8.12b}$$

$$\Sigma = 2U + S - 3\beta/2. \tag{6.8.12c}$$

$$\Xi = U + 2S - 3\beta/2. \tag{6.8.12d}$$

$$\Delta = 3U. \tag{6.8.12e}$$

$$\Sigma^* = 2U + S \tag{6.8.12f}$$

$$\Xi^* = U + 2S \tag{6.8.12g}$$

$$\Omega = 3S \tag{6.8.12h}$$

How are these obtained? For the last four states it is easy to recall the symmetry structure. Being 3/2 states they are completely symmetric in spin. So they cannot have spin = 0 components for the two body part. The first four are mixed symmetric 1/2 states and a little algebra with the Eqs. (2.1.8), (2.1.9) yields (6.8.12). One can fit the masses with $U = 412.9$, $S = 557.5$, $\alpha = 200.5$ and $\beta = 132.7$, all in MeV.

(1) The overall fit is better compared to de Rujula et al.

(2) there are three mass relations: (a) $3\Lambda + \Sigma(4543\ \text{MeV}) = 2(N + \Xi)(4512\ \text{MeV})$, – the Gell-Mann Okubo relation for octet baryons, (b) the equal spacing rule for decuplet baryons $\Xi^* - \Sigma^*(148\ \text{MeV}) = \Omega - \Xi^*(139\ \text{MeV}) = \Sigma^* - \Delta(153\ \text{MeV})$ (c) a relation between octet and decuplet baryons $\Xi^* - \Sigma^*(148\ \text{MeV}) = \Xi - \Sigma(123\ \text{MeV})$.

(3) $\beta/\alpha = 0.662$ where as $U/S = 0.74$, within 10%.

(4) The difference between U and S in previous work was 175 MeV whereas here this is only 145 MeV. This could be important when comparing with current quark masses as extracted for example by *Gasser* and *Leutwyler* (1982).

(5) One can estimate α and β from the meson sector as follows

$$\alpha = (M_\rho - M_\pi)R_\pi^3/R_N^3 = 300\ \text{MeV} \tag{6.8.13}$$

$$\beta = (M_{K^*} - M_K)R_K^3/R_N^3 = 146\ \text{MeV}$$

assuming that the instanton interaction for mesons and diquarks are the same and that the volume ratio corrects for the different probability of finding the instantons in the different systems. The order of magnitude of the parameters for the meson and the baryon cases seems to support these assumptions. For example in Eq. (13) one could get $\alpha \simeq 200$ MeV by choosing $R_\pi = 0.44$ and $R_N = 0.64$ fm. However, the important lesson, Shuryak and Rosner point out, is that one should not force agreement of baryon ground state masses by V_{OGE} with a large α_s (as done in the MIT bag model for instance) but realize that nonperturbative effects also contribute here and can explain the experimentally observed mass differences equally well.

Can one go further? The method can be applied to charm baryons as well. Recently some experimental groups investigated charmed baryons in $\bar{p}p$, e^+e^-, SpS and other colliders. So the problem of charmed baryon masses and decays is of great interest. We try the approach of Shuryak and Rosner described in the previous section.

For single charmed baryons we get (*Dey, Dey* and *Volkovitsky* 1991):

$$\Lambda_c = C + 2U - \alpha - \gamma/2 \tag{6.8.14a}$$

$$\Sigma_c = C + 2U - 3\gamma/2 \tag{6.8.14b}$$

$$\Xi_c = C + U + S - \beta - \gamma/4 - \delta/4 \tag{6.8.14c}$$

$$\Xi_c' = C + U + S - 3\gamma/4 - 3\delta/4 \tag{6.8.14d}$$

$$\Omega_c = C + 2S - 3\delta/2 \tag{6.8.14e}$$

$$\Sigma_c^* = C + 2U \tag{6.8.14f}$$

$$\Xi_c^* = C + U + S \tag{6.8.14g}$$

$$\Omega_c^* = C + 2S \tag{6.8.14h}$$

Here Ξ_c is a baryon where s and $u(d)$ quarks form a scalar diquark and Ξ_c' is a baryon with vector $u(d)$ and s diquark. It was shown in *Franklin* et al. (1981) that the mixing between Ξ_c and Ξ_c' is small. C is an effective mass of charmed quark and γ and δ are the differences of effective masses of vector and scalar diquarks combined from $u(d)c$ and sc quarks respectively. Without any further assumptions we get two relations for the charmed baryons:

$$\Xi_c' - \Sigma_c = \Omega_c - \Xi_c' \tag{6.8.15a}$$

$$\Xi_c^* - \Sigma_c^* = \Omega_c^* - \Xi_c^* \tag{6.8.15b}$$

Experimentally the masses of four charmed baryons are known:

$$\Lambda_c = 2285 \text{ MeV}, \quad \Sigma_c = 2455 \text{ MeV}, \quad \Xi_c = 2466 \text{ MeV}, \quad \Omega_c = 2740 \text{ MeV}$$

and we can predict the mass of the Ξ_c' baryon to be 2597 MeV.

If we now suppose that all parameters in Eqs. (12) and Eqs. (14) are the same (which means the effective masses and interactions of ordinary and strange quarks in charmed baryons are the same as in non-charmed baryons, – we call this the *equality* assumption), then we obtain three more relations: (1) a generalization of the Gell-Mann Okubo relation for charmed baryons with $J^P = 1/2^+$:

$$6\Lambda_c + 4\Sigma + \Omega_c = 6\Xi_c + 4N + \Sigma_c \tag{6.8.16a}$$

(2) the equal spacing rule for all $J^P = 3/2^+$ baryons:

$$\Xi_c^* - \Sigma_c^* = \Xi^* - \Sigma^* (= \Omega_c^* - \Xi_c^* = \Omega - \Xi^* = \Sigma^* - \Delta, \tag{6.8.16b}$$

(3) one new relation which involves the charmed baryons with $J^P = 1/2^+$ and $3/2^+$ and octet non-charmed baryons:

$$\Xi_c^* + 2N + \Sigma_c = 2\Omega_c + 2\Delta + 3\Lambda_c \tag{6.8.16c}$$

Eqs. (16) are weaker relations because of the equality assumption. For example, taking the experimental masses for Ξ_c, Σ_c, Λ_c, N, Σ in Eq. (16a) we find Ω_c to be only 2505 MeV. This compares very poorly with the experimental number given and we are forced to conclude that the parameters of the model for the charmed baryons must be different.

One can use a set $U = 415$ MeV, $S = 550$ MeV, $C = 1680$ MeV, $\alpha = 206$ MeV, $\beta = 163$ MeV, $\gamma = 38$ MeV and $\delta = 27$ MeV. This is quite justified in the instanton model since the parameters depend on the baryon size as pointed out in the previous section. We shall discuss this point in more detail later on. With these parameters we obtain the following values for charmed baryon masses using Eqs. (14):

$$\Lambda_c = 2285 \text{ MeV}, \quad \Sigma_c = 2453 \text{ MeV}, \quad \Xi_c = 2466 \text{ MeV}, \quad \Omega_c = 2740 \text{ MeV},$$

$$\Xi_c' = 2596 \text{ MeV}, \quad \Sigma_c^* = 2510 \text{ MeV}, \quad \Xi_c^* = 2645 \text{ MeV}, \quad \Omega_c^* = 2780 \text{ MeV}$$

$$\tag{6.8.17}$$

The ordinary baryon masses are not unreasonable with this set of parameters. We get $N = 936$ (938), $\Delta = 1245$ (1232), $\Lambda = 1095$ (1116), $\Sigma = 1136$ (1195), $\Xi = 1271$ (1318), $\Sigma^* = 1380$ (1385), $\Xi^* = 1515$ (1533) and $\Omega = 1650$ (1672), where the bracketed numbers are experimental values and all masses are in MeV. It should be noted that β and α differ by 30 and 6 MeV between our set of parameters and those of Shuryak and Rosner. Although the difference is small the factor 6 in Λ_c and Ξ_c in Eq. (11) enhances the error and lessens the Ω_c mass by 235 MeV.

As we have pointed out earlier, in the instanton model we expect the diquark binding to be inversely proportional to the volume, i.e. (using the names of the mesons for the masses)

$$(\rho - \pi)R_M^3 = (\Delta - N)R_B^3 \tag{6.8.18a}$$

and

$$(K^* - K)R_{M'}^3 = (\Sigma^* - \Sigma)R_{B'}^3. \tag{6.8.18b}$$

Here R_M and R_B with/without primes are the respective radii of the scalar/vector mesons and the octet/decuplet baryons. A value of $\alpha = 200$ MeV and $\beta = 150$ MeV was already justified by Shuryak and Rosner. Since in the bag model the volume is proportional to the mass of the hadron we would expect

$$\gamma = (3\alpha/2)(\rho + \pi)/(\Lambda_c + \Sigma_c + \Xi_c) = 38 \text{ MeV} \tag{6.8.19a}$$

$$\delta = 2\beta D/(\Xi_c + \Omega_c) = 27 \text{ MeV} \qquad\qquad (6.8.19b)$$

taking the D – meson mass to be 1869 MeV. One could of course make the equality assumption and fit C, γ and δ from the known Λ, Σ and Ω_c and predict $\Xi_c = 2497$ MeV to be compared with the experimental number 2466. This would be against the spirit of the calculation since then we find δ (34 MeV) greater than γ (32.5). We have also calculated beauty baryons of which only one Λ_b is known. In the future the charmed and beauty baryon mass relations may be put to experimental test.

7 Chiral Perturbation Theory (CHPT)

7.1 CHPT for the Meson Sector

Let us, for the sake of completeness, briefly repeat some facts which have been already stressed in this book in different contexts. The eight light hadrons (π, K, η) are pseudoscalar mesons. They are believed to be composed of the quarks (u, d, s) and their antiquarks. For yet unknown reasons, their masses (m_u, m_d, m_s) happen to be small. If these masses were strictly zero, the Lagrangian of QCD would exhibit an exact $SU(3)_R \times SU(3)_L$ symmetry. For the standard model to be consistent with the facts of life, it is crucial that the ground state of the theory is not symmetric under this group, such that this RL symmetry breaks down to $SU(3)_{R+L}$. Pseudoscalar mesons are identified with the Goldstone bosons generated by the symmetry breakdown. If m_u, m_d and m_s were zero, we would have the chiral limit, and the pseudoscalar mesons would be massless. In real world, their masses are proportional to square roots of quark masses, e.g.

$$M_\pi^2 = (m_u + m_d)B_0[1 + O(m)] \tag{7.1.1}$$

where the constant B_0 depends on the quark condensate. The hidden symmetry reveals itself in the low energy properties of the pseudoscalar mesons. *Weinberg* (1979) showed that these properties can be analyzed on the basis of an effective Lagrangian, replacing the quark and gluon fields of QCD by a meson field, represented by an element in the $SU(3)$ flavour gauge group. The effective Lagrangian is expanded in powers of derivatives of the meson field and in powers of mass matrix:

$$m = \begin{pmatrix} m_d & & \\ & m_d & \\ & & m_s \end{pmatrix} \tag{7.1.2}$$

The leading contribution is of the form

$$\mathcal{L}^{(2)} = \frac{F_0^2}{4} \langle \partial_\mu U^\dagger \partial_\mu U + 2B_0 m(U + U^\dagger) \rangle \tag{7.1.3}$$

where the symbol $\langle A \rangle$ stands for the trace of the 3$\times$3 matrix A. The constant F_0 is the value of the pion decay constant F_π in the chiral limit

$$F_0 = F_\pi[1 + O(m)] \tag{7.1.4}$$

The F_π is $\simeq 93$ MeV. The various current algebra relations are readily obtained by calculating the relevant tree graphs of $\mathcal{L}^{(2)}$. Write $U = \exp(i\varphi/F_0)$ and expand in powers of φ which is a matrix given by $\varphi = \varphi^a \lambda^a/2$. An example is the well-known low energy theorem for the decay $\eta \to 3\pi$

$$A(s,t) = \frac{\sqrt{3}}{4} \frac{m_u - m_d}{m_s - \hat{m}} \frac{s - \frac{4}{3}M_\pi^2}{F_0^2} + O(p^4) \tag{7.1.5}$$

where m is the average mass of u and d. The relation (5) is an exact statement in the following sense. The amplitude $A(s,t)$ is a function of the kinematic variables[1] s, t, of the quark masses and the scale Λ_{QCD}. The electromagnetic contributions, of order $e^2 p^2$ are disregarded. Put a scale factor λ in s, t and the quark masses and expand in powers of λ. The theorem states that there is no term of order λ^0 and the contribution of order λ^1 is determined by the constants F_0, B_0 according to Eq. (5). Alternatively one may count the powers of four-momenta. Because of the Eq. (1) and the on mass shell constraints, such as $p^2 = M_\pi^2$, the quark masses must be counted as quantities of order p^2. The low energy theorem Eq. (5) determines the amplitude $A(s,t)$ at order p^2. The corrections are of order p^4.

More generally, the T-matrix elements describing the scattering of any number of pseudoscalar mesons are of order p^2. The leading contribution, given by the tree graphs of $\mathcal{L}^{(2)}$, only involve the quark masses and the two constants F_0 B_0. Unitarity implies that at order p^4, the T-matrix is not a polynomial in momenta, but involves cuts with discontinuities determined by the square of the leading contribution to the T-matrix. In fact together with analyticity, unitarity fixes the contribution of order p^4 upto a polynomial in the external momenta.

In the language of field theory, the discontinuities required by unitarity appear in the one-loop graphs of $\mathcal{L}^{(2)}$. Formally these graphs are represented by divergent integrals which require renormalization. The renormalized graphs are unambiguous only up to a polynomial in the external momenta. The occurrence of unspecified subtraction constants in the one-loop graphs reflects the fact that $\mathcal{L}^{(2)}$ only represents the leading term in the derivative expansion

$$\mathcal{L}_{\text{eff}} = \mathcal{L}^{(2)} + \mathcal{L}^{(4)} + \mathcal{L}^{(6)} + \tag{7.1.6}$$

of the full effective Lagrangian. The contribution $\mathcal{L}^{(4)}$ of order p^4 involves seven new coupling constants which are not fixed by F_0, B_0. If one restricts oneself to T-matrix elements to order p^4, there is no contribution from the higher terms $\mathcal{L}^{(6)} \ldots$. We will not go into details but will instead go to the external field method.

If the T-matrix elements are calculated beyond leading order it is not justified to identify F_0 with F_π. To exploit the experimental information concerning the matrix elements $\langle 0|A_\mu|\pi\rangle \sim F_\pi$, it is useful to extend the frame work from a scheme which only deals with T-matrix elements to a scheme which allows one to calculate the Green's functions associated with the currents. The external

---

[1] For a scattering process $AB \Rightarrow CD$, we expect two independent kinetic variables. The Mandelstam variables in terms of four momenta are $s = (P_A + P_B)^2$, $t = (P_A - P_C)^2$ and $u = (P_A - P_D)^2$. The variables u and t correspond to processes $A\bar{D} \Rightarrow C\bar{B}$ and $\bar{D}B \Rightarrow C\bar{A}$. But $s + t + u = m_A^2 + m_B^2 + m_C^2 + m_D^2$.

field method is ideally suited for this purpose. One replaces the QCD Lagrangian by

$$\mathcal{L}_{\text{QCD}} \Rightarrow \mathcal{L}_{\text{QCD}} + \bar{q}\gamma_\mu(v_\mu + a_\mu\gamma_5)q + \bar{q}(s + ip\gamma_5)q \qquad (7.1.7)$$

where the external fields v_μ, a_μ, s and p are matrix-valued functions in flavour space but they are Lorentz vector, axial vector, scalar and pseudoscalar respectively. Green's functions of currents are obtained from vacuum transition amplitudes $\langle 0$ out $|0$ in $\rangle_{v,a,s,p}$ by expanding in powers of the external field (Chapter 4.4).

At low energies the transition amplitude can be analyzed in the same manner as the T-matrix elements. The only difference is that, now the effective Lagrangian not only contains the meson fields $U(x)$, but also the external fields. In the effective meson theory, the vacuum transition amplitude is given by the functional integral

$$\langle 0 \text{ out } |0 \text{ in } \rangle_{v,a,s,p} = \int [dU]\exp\{\int dx \mathcal{L}_{\text{eff}}(U, v, a, s, p)\} \qquad (7.1.8)$$

In the leading term $\mathcal{L}^{(2)}$, the modification brought about by the external fields is very simple. It suffices to replace the ordinary derivative by the covariant derivative

$$\mathcal{D}_\mu U = \partial_\mu U - i(v_\mu + a_\mu)U + iU(v_\mu - a_\mu) \qquad (7.1.9)$$

and to replace the quark mass matrix by the external field $s(x)+ip(x)$. If we define $\chi(x) \equiv 2B_0[s(x) + ip(x)]$, we get

$$\mathcal{L}^{(2)} = \frac{F_0^2}{4}\langle \mathcal{D}_\mu U^\dagger \mathcal{D}_\mu U + \chi^\dagger U + U^\dagger\chi\rangle \qquad (7.1.10)$$

Next we shall discuss briefly the p^4 terms and refer to the original paper (*Gasser* and *Leutwyler*, 1985) without listing their complicated forms. In the evaluation of $\mathcal{L}^{(4)}$ the field $U(x)$ is needed only in lowest order; one may therefore use the equation of motion associated with $\mathcal{L}^{(2)}$:

$$\mathcal{D}^\mu\mathcal{D}_\mu UU^\dagger - U\mathcal{D}^\mu\mathcal{D}_\mu U^\dagger + U\chi^\dagger - \chi U^\dagger + \frac{1}{3}\langle U\chi^\dagger - \chi U^\dagger\rangle = 0 \quad (7.1.11)$$

The number of independent chiral invariants can be reduced by making use of this relation. Disregarding total divergences there are still 12 independent terms of order p^4. These terms can be divided into 3 distinct species: 7 on-shell couplings, 3 off-shell couplings and 2 contact terms. They are not all accessible to direct experiment. Three off-shell terms determine the ratio F_K/F_π, the meson charge radii and amplitude of the decay $\pi \longrightarrow ev\gamma$; conversely the experimental information on these quantities allows one to determine the values of the off-shell coupling constants within small errors. One of the contact terms control $\langle 0|\bar{s}s|0\rangle/\langle 0|\bar{u}u|0\rangle$ for which there is some indirect information from experiment. Since the coupling constants for the twelve terms renormalize the logarithmic divergences occurring in the one-loop graphs, their value not only depends on the parameters $\Lambda_{\text{QCD}}, m_c, m_b$ etc., but they also depend on the scale μ at which one-loop graphs are renormalized. In the large N_c limit the scale dependence disappears, because the one loop graphs are proportional to $1/F^2$ which goes as $1/N_c$. In the large N_c limit all except one of the 12 couplings are pure numbers. The exceptional one receives a contribution from η' exchange of order N_c^2. Notice that all results of CHPT contain error bars, due to the error bars in the evaluation of the renormalized constants which have to be taken from experiment.

A quite different estimate of the coupling constants is obtained if the cutoff needed to regularize the one-loop graphs is taken seriously. Cutting the loop integrals off at p_{max}, and assuming that no extra contributions involving higher order terms in the derivative expansion are needed if the cutoff is chosen properly, one obtains a representation of the various amplitudes which involves a single parameter p_{max} instead of the *fauna* of coupling constants described above. The prescription is simple; unfortunately it is quite wrong. A case in point is the pion charge radius. The corresponding one-loop contribution is logarithmically divergent. With a cutoff of the order of $p_{max} \approx 1$ GeV, one obtains $\langle r^2 \rangle_\pi \simeq 0.09 fm^2$. An inflation of the cutoff by a factor 2 increases the result by $0.06 fm^2$. Only an absurd value like $p_{max} \approx 50$ GeV would produce a charge radius which resembles the experimental result $\langle r^2 \rangle_\pi = 0.44 fm^2$.

Before discussing some of the results obtained on the basis of CHPT, two remarks concerning the technology are in order. One issue is the role of the pion field and its renormalization. The external field method solves this problem by dismissal: the formula (8) shows that, in this framework, the pion field is a mere variable of integration devoid of physical significance. The external field method does not involve Green's functions associated with a Hermitean operator $\varphi(x)$ or with a unitary operator $U(x)$. Instead, the method unambiguously specifies the Green's functions of the pseudoscalar operator $\bar{q} \lambda^i \gamma_5 q$; in particular, it determines the corresponding field renormalization constant G_π

$$\langle 0 | i \bar{q} \lambda^i \gamma_5 q / 2 | \pi_k \rangle = \delta_k^i G_\pi. \tag{7.1.12}$$

To leading order in the energy expansion, the operator $\bar{q} \lambda^i \gamma_5 q$ coincides with the classical field $F_0^2 B_0 \, Im \, tr \, \lambda^i U(x)$. The structure of the effective Lagrangian shows, however, that at order p^2, the representation of this operator in terms of $U(x)$ picks up an additional contribution. There is no decent pion field operator, because there is no counterpart to the variable $U(x)$ in QCD.

The other point refers to regularization schemes. Dimensional regularization is the scheme adopted, as it not only respects Lorentz invariance and chiral symmetry, but it also protects the coupling constants F_0, B_0 from renormalization.

The first application of CHPT we describe is the analysis of the matrix element $\langle K | V^\mu | \pi \rangle$. *Leutwyler* and *Roos* (1984) determine the Cabbibo angle V_{us} to be $0.220 \pm .002$, (which is in agreement with semileptonic hyperon decay according to *Donoghue, Holstein* and *Klimt*, 1984), and get $F_K / F_\pi = 1.22 \pm .01$. Another example is $\pi \pi$ scattering lengths which show significant difference from the low energy theorems of current algebra. Loop corrections remove the differences (*Gasser* and *Leutwyler*, 1984) except the case of the $I = 0$, S-wave scattering lengths for which they get $a_0^0 = 0.20 \pm .01$ to be compared to current algebra value .16 and the number extracted from experiment: $.26 \pm .05$. There are discrepancies in some other cases too. For the scalar form factor $\langle K | \bar{u} s | \pi \rangle$, *Gasser* and *Leutwyler* (1985) obtain a low energy theorem relating the slope λ_0 to the ratio F_K / F_π.

With the observed value of F_K/F_π as an input CHPT predicts $\lambda_0 = .017 \pm .001$ whereas recent experiments (*D. Hill* et al. 1979, *Y. Cho* et al. 1980, *V. Birulev* et al. 1981) report much higher values ≈ 0.046. All would be well if the high statistics old SLAC result (*Donaldson* et al. 1974) $\approx 0.019 \pm .004$ is recovered back in future experiments. Finally the decay $\eta \longrightarrow 3\pi$ which we started with is also in difficulty. With large one-loop corrections CHPT gives the value for the width $\Gamma_{\eta \to 3\pi} = 160 \pm 50$ eV whereas the recent experimental result is ≈ 317 eV. The nucleon sector, which we discuss next, is also plagued with such problems when CHPT faces experiment.

7.2   CHPT for the Nucleon

Since the original papers on CHPT are voluminous, we have to discuss it very briefly. This applies to the previous sector as well as the present one, the pioneering paper being *Gasser, Sainio* and *Svarc* (1988, GSS in short). At the very start we should note the way GSS introduce the nucleon: through external fields like the vector and other fields. This is not the only way this can be done and possibly, in future other groups will explore ways of doing things which is different and, may be complementary. One such approach already exists, the Skyrme model: there the nucleon is a topological object in the U field.

The GSS introduce an extra Lagrangian density to the one in Eq. (1.6):

$$\mathcal{L}_1 = \mathcal{L}_{\pi N}^{(1)} + \mathcal{L}_2 + \bar{\psi}\eta + \bar{\eta}\psi \tag{7.2.1}$$

with η and $\bar{\eta}$ not to be confused with the η meson of the last section. They are Grassmann variables which must accompany fermion fields since the latter occur linearly in $\mathcal{L}$.

The $\mathcal{L}_{\pi N}^{(1)}$ is simple, involving $\mathcal{D}$ defined in Eq. (1.9) and u defined through $u^2(x) \equiv U(x)$:

$$\mathcal{L}_{\pi N}^{(1)} = -\bar{\psi} D_1 \psi,$$

$$D_1 \equiv -i\gamma^\mu \mathcal{D}_\mu + M - (i/2)g_A \gamma^\mu \gamma^5 u^\dagger \mathcal{D}_\mu U u^\dagger \tag{7.2.2}$$

but the correction terms $\mathcal{L}_2$ contain further πN interaction terms and we will not write them out in detail. The approximate nucleon propagator can be written out now from (2.2) as

$$S^{(1)}(p', p) = -\int \mathrm{d}x\mathrm{d}y e^{i(p'x - py)} \frac{\delta}{\delta\bar{\eta}(x)} \frac{\delta}{\delta\eta(y)} \mathcal{L}_1$$

$$= d_0^{-1} - d_0^{-1}d_1 d_0^{-1} - d_0^{-1}d_2 d_0^{-1} + \ldots. \tag{7.2.3}$$

where

$$d_0 = -i\gamma^\mu \partial_\mu + M; \quad d_2 = -i\gamma^\mu \Gamma_\mu \text{ and}$$

$$2d_1 = -ig_A \gamma^\mu \gamma^5 u^\dagger \mathcal{D}_\mu U u^\dagger$$

$$\Gamma_{\mu} = \frac{1}{2}[u^{\dagger}, \partial_{\mu}u] - \frac{1}{2}iu^{\dagger}(v_{\mu} + a_{\mu})u - \frac{1}{2}iu(v_{\mu} - a_{\mu})u^{\dagger}. \tag{7.2.4}$$

M is the nucleon mass in the above formulae. The complicated expressions above show that not only the pion-nucleon but the interactions of the nucleon with the vector (axial vector) mesons are also included here. The nucleon propagator has been applied to the calculation of the so-called σ term, which is proportional to the expectation value of $(\bar{u}u + \bar{d}d)$ in the proton. This is related to even-isospin πN scattering amplitude Σ, by $\Sigma - \sigma \simeq 5$ MeV. The quantity one calculates in CHPT is σ_0 which turns out to be 35 ± 5 MeV. This is related to σ through a parameter y as $\sigma = \sigma_0/(1 - y)$. Present day experiment gives Σ to be 64 ± 8 MeV. Using this we get

$$y \equiv \frac{2\langle p|\bar{s}s|p\rangle}{\langle p|\bar{u}u + \bar{d}d|p\rangle} \simeq 0.4 \tag{7.2.5}$$

which implies a 40% component of ss in the proton wavefunction $|p\rangle$. Later *Gasser* et al. (1991) jacked this down to 20% by reestimating sigma from CHPT, which gives isoscalar charge radius larger than the electromagnetic one. This has also been questioned by *S. Coon* and *M. Scadron* (*Baryon* 1992 preprint). Even 20% $\bar{s}s$ in proton does not seem to be acceptable at the present by the nuclear physics community. New precision experiments are planned and hopefully these will resolve the crisis.

7.3 CHPT at Finite Temperature

In last two sections we have described how experiment helps chiral perturbation theory to fix up its renormalization constants and then it can be compared to other experimental data. It has been extended to finite temperature by *Gerber* and *Leutwyler* (1989). Analogous to Eq. (6.5.9) we can write

$$Tr\, e^{-\beta H} = \int D[U]\, e^{-\int \mathcal{L}_{\text{eff}} d^4 x} \tag{7.3.1}$$

where the integration extends over all pion field configurations which are periodic in the Euclidean time direction $U(\underline{x}, x_4 + \beta) = U(\underline{x}, x_4)$. Restricting oneself to the infinite volume limit, with box periodic conditions in a cube L^3, we can define the partition function Z as:

$$Z = -T \lim_{L \to \infty} L^{-3} \ln[Tr \exp(-H/T)] \tag{7.3.2}$$

If the temperature is low compared to the pion mass, the partition is dominated by the contribution from the ground state. Since the system is homogeneous the pressure P is given by $P = \varepsilon_0 - Z$. At low temperatures, P is of the order of $\exp(-M_{\pi}/T)$, as the leading contribution stems from one pion states, whereas states containing two or more pions show up at order $\exp(-2M_{\pi}T)$. It is therefore not necessary to evaluate the pion self-energy in a separate calculation. We can read off this value of the physical pion mass from the behaviour of the pressure at low temperatures:

$$M_\pi = - \lim_{T \to 0} T (\ln P) \tag{7.3.3}$$

The most important result of CHPT at finite T is the prediction of a first order phase transition where for example the quark condensate melts down with a temperature dependent factor $x = T^2/8F^2$ ($F = 88$ MeV) as follows

$$\langle qq \rangle_T = \langle qq \rangle_0 [1 - x - x^2/6 \ldots] \tag{7.3.4}$$

Unfortunately the recent lattice calculations do not support such a picture in the presence of two light flavours. However, based on current algebra wisdom, CHPT must be valid at asymptotically low temperatures. So it is a good starting point for investigating the mysterious lattice results found recently by the relevant people.

8 The Topological and Non-Topological Soliton Models

8.1 Skyrmion: Baryon as a Topological Soliton in Meson Fields

We have already talked about large N_c expansion of 't Hooft and Witten's application to baryons. Witten had, in particular, observed that the mass of baryons go as $1/g^2 (\approx N_c)$ and this looks remarkably like topological Hamiltonian or action terms discussed in our Chapters 5 and 6. Could baryons be described as solitons in meson fields? The answer was affirmative. One only had to *go back 30 years* to Skyrme's papers of 1960-s. *Balachandran* et al. (1982, 1983) had already revived these papers. So *Adkins, Nappi* and *Witten* (1983) reformulated Skyrme's solutions for N and Δ. This was also done by *Jackson* and *Rho* (1983). Since then there has been lot of work on Skyrme model and it is reviewed in *Zahed* and *Brown* (1986) and in Bhaduri's book (1988). After 1988 there have been two interesting new developments. One of these is the connection between the Skyrmion and the instanton, already discussed in Chapter 6. We will describe the Skyrme Lagrangian and then refer to the other work: quantum stabilization of the Skyrme soliton.

The Skyrme Lagrangian density is

$$\mathcal{L} = \frac{F_\pi^2}{4} Tr(\partial_\mu U^\dagger \partial_\mu U) + \frac{\varepsilon^2}{4} Tr[U^\dagger \partial_\mu U, U^\dagger \partial_\nu U]^2$$

$$= \frac{F_\pi^2}{4} Tr(L_\mu^2) + \frac{\varepsilon^2}{4} Tr[L_\mu^\dagger, L_\nu]^2 \qquad (8.1.1)$$

where $L_\mu \equiv U^\dagger \partial_\mu U$. There is a conserved current (topological as opposed to a Noether current because it is not associated with any property of $\mathcal{L}$): $B_\mu = (1/24\pi^2)\varepsilon_{\mu\nu\delta\lambda}L_\nu L_\delta L_\lambda$ which is conserved, $\partial_\mu B_\mu = 0$ and the fourth component of this current integrated over space gives us the baryonic number. Notice that the first term is just like the Eq. (7.1.3) of chiral perturbation theory (CHPT). Here $F_\pi \simeq 93$ MeV and ε is dimensionless. The latter is sometimes written as $1/8g_\rho^2$ (*Bhaduri*, 1988), where $g_\rho \simeq 5.45$ is intriguingly close to the universal coupling of the ρ-meson given by vector dominance. This in fact has been exploited to establish a connection with vector dominance and the Skyrme model, discussed in detail in Chapter 7 of *Bhaduri*, 1988. In a more recent work *Novozhilov* (1989) suggests that one should use $F_\pi \simeq 88$ MeV, and then use the chiral logarithm corrections of CHPT. He suggests that the ρ-meson coupling g_ρ should have the value 2π, curiously close to the value found by Skyrme from consideration of spin quantization. It seems this value remains invariant with change in temperature and density (for a discussion see *Dey, Dey* and *Ghose*, 1987).

One of the problems with the Skyrme model is that it has hedgehog solutions for spin and isospin coupled up, so that $\underline{K} = \underline{S} + \underline{T}$, is a good quantum number instead of both $\underline{S}$ and $\underline{T}$. One has to project good angular momentum states and although it leads to sensible results for the ground states, it is not easy to explain excited states with $K \neq 0$, i.e. with different S and T.

The results of the Skyrme model are not particularly good as can be seen from the last column of Table 8.1. Still it has attracted much attention probably because of its novelty. One of the problems is that the particular form of the quartic term is hard to explain. The straight forward loop expansion of the pion Lagrangian leads to divergent by renormalizable terms of CHPT, so how does one justify the quartic term? One of the alternatives is to consider a composite ρ-meson field, and then the kinetic energy of this reduces to the quartic term. This is the basic idea behind the hidden symmetry models we already referred to.

The other alternative is to do away with quartic term altogether. But the quartic term is responsible for stabilization of the Skyrmion in the original Skyrme model. Without it how can one get stability? The answer is to allow quantum vibration in the radius R of the unstable soliton. For small R, now a function $R(t)$, the quantum kinetic energy is large and is enough to stabilize the soliton as was shown by *Jain* et al. (1989). The quantum vibrations were also considered before by *Carlson* (1985) and by *Mignaco* and *Wulck* (1989). We have given the results of Jain, incorporating the correction reported by *Bhaduri* et al. (1990) and it may be noticed that in fact the results of the new Skyrme model in the above Table (8.1) are as good as the old, may be better. The problem is that there is no *a priori* justification for including quantum corrections for a particular collective mode, leaving out others. So it is possible that the results are good *accidentally*, there being no physical basis for it. However, given the choice of $R(t)$ as a collective mode, along with angular momentum projection, the numerical results of Jain can be obtained in a simple variational model (*Bhaduri* et al. 1990).

Table 8.1. Results of the Skyrme model with and without the quartic term

	Experiment	Without quartic (*Jain* et al. 1989)	With quartic (*Jackson* and *Rho* 1983)
M_N (GeV)	0.939	1.14	1.34
M_Δ (GeV)	1.232	1.46	1.77
$\langle r_0^2 \rangle$ (fm)	0.79	0.58	0.41
μ_N	0.88	0.64	0.55
$\mu_{\Delta N}$	4.70	3.01	3.09
g_A	1.23	0.98	0.60

8.2 The Non-topological Soliton or the Soliton Bag Model

Another attractive model of QCD, the soliton bag, was introduced originally by *Huang* and *Stump* (1976) and further developed by *Friedberg* and *Lee* (1977, 1978). It is a covariant field theory and sufficiently general so that, for certain adjustable parameters, it can describe the MIT bag of Section 2.2. But generally it is a more powerful model, particularly the versions known as colour dielectric models, where total confinement can be achieved. The earlier models were also useful and were extensively studied: tricky issues like centre of mass subtraction and boost invariance were sorted out. For a review we refer the reader to the monograph by *Wilets* (1989) and the review article by *Birse* (1990) where extensive references may also be found. We will refer to recent work reported in the last two years. Next we move on to the non-topological solitons models where the quark fields are not completely left out.

8.3 The Nature of the QCD Vacuum

In QED, the vacuum behaves as a polarizable medium with the dielectric constant $K > 1$. If there is a charge distribution, the virtual cloud of electron-positron pairs created by virtual photons will screen the charge as long as the distance r is large enough. In a Lorentz invariant theory, $\mu K = 1$ where the permeability $\mu < 1$ as $K > 1$. In other words, the medium is diamagnetic. A nice discussion on this is given in the book by *Lee* (1981) and *Bhaduri* (1988).

In QCD gluons play the same role as photons. Quarks interact via gluons. As a result, the QCD vacuum also should be diamagnetic except for the fact that the gluons themselves have colour charges as we have discussed right at the start (Chapter 1). Colour $\mu_c > 1$ as the gluons produce magnetization of the medium and therefore $K_c < 1$, where the subscript stands for colour. This makes the QCD vacuum paramagnetic. Since $K_c < 1$, the colour dielectric interaction increases with distance, producing *antiscreening*. If there is a colour-charge distribution, however small, the medium will develop a hole around the charge distribution. Inside the hole $K_c = 1$, and outside $K_c \ll 1 \cong 0$. To make the hole disappear we need infinite energy as it means compression of charges of equal sign. *Lee* (1981) has shown that a colour singlet hole has finite energy.

8.4 Description of the Model

An effective scalar field σ describes the complex structure of the QCD vacuum arising from gluon and quark condensates in this model, with a potential $U(\sigma)$

which is typically quartic in σ:

$$U(\sigma) = B + a\sigma^2/2! + b\sigma^3! + c\sigma^4/4! \tag{8.4.1}$$

In the original Friedberg–Lee model this σ field was coupled to the quark field, violating chiral symmetry. There are two ways to get a chirally symmetric model. We will first describe the model in which the σ field is coupled to the quarks indirectly through a vector gluon field (*Fai*, *Perry* and *Wilets* 1988, *Krein* et al. 1991). We refer to the recent calculation: the Lagrangian is

$$\mathcal{L} = \bar{q}(i\gamma^\mu \partial_\mu + g_s \frac{\lambda^a}{2} A_\mu^a \gamma^\mu)q - \frac{1}{4}\chi(\sigma)F_{\mu\nu}^a F^{a\mu\nu}$$

$$+ \frac{1}{2}(\partial_\mu \sigma)^2 - U(\sigma) + \text{ counterterms} \ldots \tag{8.4.2}$$

with $\chi(\sigma) = 1 + \theta(\chi)\chi^3(3\chi - 4)$, with $\chi = \sigma/\sigma_v$, and B determined by the condition $U(\sigma_v) = 0$. The parameters found were $a = 0$, $b = 400$ fm^{-4}, $c = 5000$ and the radius came out to be 0.83 fm. The "pion quark" coupling $u(r)v(r)m(r)$ was surface-peaked. The model is not yet refined enough to compare with experiment.

The second way to make the soliton model chirally invariant is to introduce pion fields $\underline{\phi}$ and χ fields explicitly. The Lagrangian is

$$\mathcal{L} = \bar{q}i\gamma_\mu \partial_\mu q + \frac{1}{2}\left[(\partial_\mu \chi)^2 - \frac{g\bar{q}(\sigma + i\tau \cdot \underline{\phi})q}{\chi}\right] - U(\sigma, \underline{\phi}) +$$

$$+ \frac{1}{2}(\partial_\mu \sigma)^2 + \frac{1}{2}(\partial_\mu \underline{\phi})^2 \tag{8.4.3}$$

where $U(\sigma, \underline{\phi}) = F_\pi m_\pi^2 \sigma + \lambda^2/4(\sigma^2 + \underline{\phi}^2 - v^2)^2$, the first term on the right being the symmetry breaking term. On projecting good linear and angular momentum one gets the magnetic moments and radii of the neutron and proton. An additional one gluon exchange is necessary to get the N-Δ splitting right, this was missing in a similar calculation by *Fiolhais* et al. (1991). Both the N and the Δ have larger masses but other data compare extremely well with experiment (*Leech* and *Birse*, 1992).

9 QCD Sum Rules

9.1 Introduction to QCD Sum Rules

The coupling constant g in the gluon tensor $G_{\mu\nu}^a = \partial_\mu A_\nu^a - \partial_\nu A_\mu^a + g f^{abc} A_\mu^b A_\nu^c$ is a bare one. Actually, all physical processes are described by the running or effective coupling $\alpha_s(Q^2) \equiv g^2(Q^2)/4\pi$, characterizing interactions at momenta Q^2, where $Q^2 = -q^2$ (at distances $r \approx Q^{-1}$). The asymptotic freedom (*Politzer*, 1973; *Gross* and *Wilczek*, 1973) takes place for $\alpha_s(Q^2) \approx 2\pi/(9\ln Q/\Lambda)$ (for 3 flavours). Because of logarithmic fall off of the coupling the structure of the theory at short distance is simple; the dynamical analysis can be carried out in terms of quarks and gluons interacting perturbatively. Here more or less the same logarithims occur as in electrodynamics. The genuine hadronic theory starts at a length of 0.5 fm and larger.

At such distances non-perturbative effects play a key role; the effective coupling constant α_s is large. *Shifman, Vainstein* and *Zakharov* (SVZ in short, 1979) started a method, called QCD sum rule (QSR, in short) method to address this region. This is based on first principle QCD and allows one to relate phenomenological information on the nontrivial vacuum structure of QCD. The idea of QSR is to approach the bound state problem in QCD from the asymptotic freedom side, i.e. to start at short distances and move to larger ones where confinement effects become important. Asymptotic freedom starts to break down and resonances emerge as a reflection of the fact that quarks and gluons are permanently confined within hadrons. The breakdown of asymptotic freedom is signalled by the introduction of non-vanishing vacuum expectation values of quark and gluon condensate operators. The general idea of the QCD vacuum is that it is densely populated by fluctuating fields whose amplitude is so large that they cannot be treated by perturbation theory. The strength of the fluctuations is characterized on average by vacuum condensates; the most important of them are quark and gluon condensates

$$\langle 0 | \frac{\alpha_s}{\pi} G_{\mu\nu}^a G_{\mu\nu}^a | 0 \rangle; \qquad \langle 0 | \bar{q}q | 0 \rangle \tag{9.1.1}$$

A pair of quarks "injected" in the vacuum evolves not in empty space but in "vacuum media". As long as the interquark distance is not large, their dynamics is determined by averaged vacuum characteristics, as in (1) and can be handled theoretically. The guess is that this theoretically reliable domain can be stretched up to distances at which classical hadrons such as the ρ-mesons etc. are formed.

In simple terms the method uses the simplicity of QCD at large momentum transfer where asymptotic freedom makes it possible to use perturbation theory. This is then extrapolated to extract information about bound state systems through non-zero averages of vacuum fields (Eq. (1)). One cannot prove that the only states are colour singlets but once the assumption is made the wealth of predictions is rewarding. The spectrum of applications of QSR is very broad. *Actually, there are not so many problems in low-energy hadron physics not studied in this framework.* Here we shall describe the method and some of its applications relevant to low energy hadron physics. We refer to the review articles by *Reinders, Rubinstein* and *Yazaki* (1985); and books by *Pascual* and *Tarrach* (1984) and *Narison* (1989).

We start with a current, a quark bilinear to start with which will be later generalized to more complicated baryon currents. One constructs a correlator from this current. Let $\bar{q}_i \Gamma q_j = J_\Gamma$, be the current which is associated with the meson angular momentum channels J^{PC} (P = parity, C = Charge conjugation) as follows:

$J_a = \bar{q}_i q_j$, $J^{PC} = 0^{++}$, scalar

$J_p = \bar{q}_i \gamma_5 q_j$, $J^{PC} = 0^{-+}$, pseudoscalar

$J_v = \bar{q}_i \gamma_\mu q_j$, $J^{PC} = 1^{--}$, vector

$J_A = \bar{q}_i \gamma_\mu \gamma_5 q_j$, $J^{PC} = 1^{++}$, axial vector and then tensor etc.

The vacuum polarization induced by such currents is given by the correlation function

$$T_{\mu\nu..}\Pi^j(Q^2) = i \int d^4x e^{iqx} \langle 0|T[j_\Gamma(x)\bar{j}_\Gamma(0)]|0\rangle \tag{9.1.2}$$

where $\Gamma = 1, \gamma_5, \gamma_\mu, \gamma_5\gamma_\mu$ etc. $\Pi^j(Q^2)$ is a scalar function of Q^2, $T_{\mu\nu..}$ a tensor depending on Γ and T on the right-hand side denotes time-ordering.

On grounds of causality and the analyticity of Π one can write a dispersion relation connecting it with its imaginary part (see for example, *Bjorken* and *Drell*, 1965)

$$\Pi^j(Q^2) = \frac{q^{2n}}{\pi} \int \frac{\text{Im}\,\Pi^j(s)ds}{s^n(s-q^2)} + \sum_{k=1}^{n-1} a_k(q^2)^k \tag{9.1.3}$$

where n is the number of subtractions necessary to make the expression *analytic* in case it has poles. The unknown subtraction constants can be eliminated by taking the appropriate number of derivatives of Π with respect to Q. Im Π is in turn related to a cross section, for example for the vector current,

$$\text{Im}\,\Pi^V(s) = \frac{9}{64\pi^2\alpha^2}s\sigma(e^+e^- \rightarrow \text{hadrons}) \tag{9.1.4}$$

where $\alpha = 1/137$ is the fine structure constant and s is the Mandelstam variable we defined in Chapter 7. By selecting a particular flavour, e.g. charm in the case of

$j_\mu(x) = \bar{c}\gamma_\mu c$, $J/\psi, \psi', \psi'', \ldots$ and continuum states above threshold ($\bar{D}D$ etc.) appear with $J^{PC} = 1^{--}$ in $\operatorname{Im} \Pi^V$. Similarly one can pick up other flavours and other currents. Thus, corresponding to a particular current j_Γ one can connect the hadronic states with a continuum through the Eqs. (2) and (4).

To parametrize $\operatorname{Im} \Pi^j(s)$ in the hadron sector the following observations are important. In general the imaginary part of the amplitudes are rapidly saturated by a few poles – due to the large universal Regge slope of these amplitudes studied extensively in the 50-s and the 60-s. This is also explained by the large string tension, or the slope of the confinement potential if one believes in such a thing. A large string tension is at par with the experimental fact that the meson and baryon resonances are far apart. But to use this property one has to essentially presuppose the spectrum. For example, in the vector current of flavour q with charge e_q, one writes

$$\operatorname{Im} \Pi^V(s) = \frac{\pi}{e_q^2} \sum \frac{m_R^2}{g_R^2} \delta(s - m_R^2) + \frac{1}{4\pi}\left(1 + \frac{\alpha_s}{\pi}\right)\theta(s - s_0) \qquad (9.1.5)$$

The θ-function on the right stands for the continuum starting at s_0 and the summation is over the resonances with masses m_R. In the vector case g_R is related to the electronic width of the resonance but for other currents the coupling has no direct physical meaning.

So far there is no connection with QCD. Only the phenomenological side of (2) has been written in terms of resonance masses and couplings (m_R and g_R). This has to be related to the correlation function obtained from QCD. This is the theoretical side which is very difficult to get because of its non-linear and non-perturbative nature. SVZ circumvent this problem by assuming that the operator product expansion holds even in the nonperturbative sector.

9.2 The Operator Product Expansion (OPE)

The T ordered product of currents (Eq. (1.2)) is expanded in terms of operators of different dimensions (*Wilson*, 1969). This expansion, proved to all orders in perturbation theory, is valid at short distances, i.e. $Q^2 = -q^2$ large:

$$i \int d^4x e^{iqx} T(A(x), B(0)) = \sum_n C_n^\Gamma(q^2) O_n \qquad (9.2.1)$$

The equation says: the time ordered product of two local operators A and B (currents in this case) may be expanded at short distances in terms of local operators O_n. $C_n^\Gamma(q^2)$ are Wilson coefficients and are functions of q^2. They depend on the Lorentz indices and quantum numbers of j_Γ and O_n. The operators are ordered by dimension and the C_n^Γ fall off by corresponding powers of q. First term of the sum in (1) is an identity operator and the corresponding Wilson coefficient C_I is the perturbative contribution (Fig. 9.1) which is logarithmically divergent. All the other coefficients are power corrections to C_I. The operators concerned here are

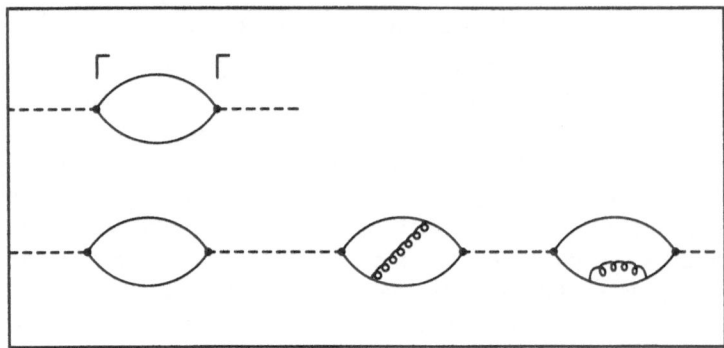

Fig. 9.1. Diagrams contributing to the unit operator (Eq. (9.2.1)) upto first order in α_s. Curly lines depict gluons, Solid lines quarks and the dashed line, the currents. These are logarithmically divergent diagrams and correspond for example, to the first term in Eq. (9.3.1) in the vector channel

those of vacuum fields (Eq. (1.1)). Thus the OPE factorizes short and large distances. The coefficients are calculated in perturbation theory by Feynman diagram techniques, while all large distance effects are buried in the matrix elements of operators O_n. For $\Pi^j(Q^2)$ the vacuum expectation values are required and since the vacuum is scalar we need the scalar operators. The complete set of such operators of dimension $d \leq 6$ is given below:

$$I \text{ (unit operator)} \qquad\qquad d = 0$$

$$O_m = m\bar{q}q \qquad\qquad d = 4$$

$$O_G = G^a_{\mu\nu}G^a_{\mu\nu} \qquad\qquad d = 4$$

$$O_\Gamma = \bar{q}\Gamma_1 q\bar{q}\Gamma_2 q \qquad\qquad d = 6$$

$$O_\sigma = m\bar{q}\sigma_{\mu\nu}\frac{\lambda^a}{2}qG^a_{\mu\nu} \qquad\qquad d = 6$$

$$O_f = f_{abc}G^a_{\mu\nu}G^b_{\nu\lambda}G^c_{\lambda\mu} \qquad\qquad d = 6 \qquad\qquad (9.2.2)$$

where m is the quark mass, the λ^a matrices are given in Chapter 2 and $\sigma_{\mu\nu} = {}^i\!/_2\left[\gamma_\mu, \gamma_\nu\right]$. Explicitly, using the sum (1) and taking the vacuum matrix elements we have

$$\Pi^j(Q^2) = C_I + C_G \langle 0|\frac{\alpha_s}{\pi}G^a_{\mu\nu}G^a_{\mu\nu}|0\rangle + C_m \langle 0|m\bar{q}q|0\rangle + \dots . \qquad (9.2.3)$$

This is the theoretical side of the sum rule. Equating this with Eqs. (1.3–1.5) we get relations between the parameters of the theory and those of hadrons.

The power correction terms ($C_G, C_m \dots$) are very important. Consider the vacuum polarizations (Π) of the vector current $j^V_\mu(x) = \bar{q}\gamma_\mu q$ and the axial vector current $j^A_\mu(x) = \bar{q}\gamma_\mu\gamma_5 q$. In the chiral limit $m_q \Rightarrow 0$ and the perturbative diagrams (Fig. 9.1) do not differentiate between the coupling $\Gamma = \gamma_\mu$ or $\Gamma = \gamma_\mu\gamma_5$, γ_5 in the quark-current vertex of Π^A can be pulled through to the other side without a sign change so that Π^A reduces to Π^V. In the real world, however, the masses of ρ and A_1 are very different and pion is almost massless. In other words, the condensates contribute non-trivially towards the $\pi - \rho - A_1$ mass splitting.

9.3 Calculation of the Coefficients and Borel Transform

The coefficients C in (2.3) are calculated in deep Euclidean region ($Q^2 = -q^2$) where asymptotic freedom prevails and perturbation theory can be used. Consider the iso-vector, vector channel in the light sector, namely the resonance ρ. The current is given by $j_\mu(x) = \frac{1}{2}(\bar{u}\gamma_\mu u - \bar{d}\gamma_\mu d)$. The correlator under consideration is a two point function. There are three relevant mass parameters. The first one is the inverse of the confinement radius, $R_{con}^{-1} \cong \mu$ which is manifested (for example) through the expectation value of $\langle 0|\alpha_s/\pi G_{\mu\nu}^a G_{\mu\nu}^a|0\rangle \simeq \mu^4$. The second is the quark mass m_q and the third one is the external mass scale introduced by the momentum q in the definition of the two point function (Eq. (2.1)). Since $m_q \lesssim \mu$ and $Q^2 \gg \mu^2$ the expansion parameters in this sector are μ^2/Q^2 and m_q^2/Q^2. One gets

$$i\int d^4 x e^{iqx} \langle 0|T[j_\mu(x)\bar{j}_\nu(0)]|0\rangle = (q_\mu q_\nu - q^2 g_{\mu\nu})\Pi(Q^2)$$

where

$$\Pi(Q^2) = -\frac{1}{8\pi^2}\left(1 + \frac{\alpha_s}{\pi}\right)\ln\frac{Q^2}{\mu^2} + \frac{1}{2Q^4}\langle 0|m_u\bar{u}u + m_d\bar{d}d)|0\rangle$$

$$+ \frac{1}{24Q^4}\langle 0|\frac{\alpha_s}{\pi}G_{\mu\nu}^a G_{\mu\nu}^a|0\rangle - \frac{\pi\alpha_s}{2Q^6}\langle 0|\bar{u}\gamma_\mu\gamma_5\lambda^a u - \bar{d}\gamma_\mu\gamma_5\lambda^a d)^2|0\rangle$$

$$+ \frac{\pi\alpha_s}{9Q^6}\langle 0|(\bar{u}\gamma_\mu\lambda^a u + \bar{d}\gamma_\mu\lambda^a d)\sum_{q=u,d,s}\bar{q}\gamma_\mu\lambda^a q|0\rangle \qquad (9.3.1)$$

In this Eq. the first term is the perturbative contribution with the infra-red cut-off μ^2. The other terms are the power corrections (Fig. 9.2a,b). The last two terms are due to four fermion operators O_Γ. They correspond to diagrams (Fig. 9.2c) where large momentum flows through either an internal gluon line or quark line. SVZ saturated the intermediate states by the vacuum and found

$$\langle 0|(\bar{q}\gamma_\mu\lambda^a q\bar{q}\gamma_\mu\lambda^a q|0\rangle = -\frac{16}{9}\langle 0|\bar{q}q|0\rangle^2$$

$$\langle 0|(\bar{q}\gamma_5\lambda^a q\bar{q}\gamma_5\lambda^a q|0\rangle = -\frac{4}{9}\langle 0|\bar{q}q|0\rangle^2 \qquad (9.3.2)$$

so that

$$\Pi(Q^2) = -\frac{1}{8\pi^2}(1 + \frac{\alpha_s}{\pi})\ln\frac{Q^2}{\mu^2} + \frac{1}{2Q^4}\langle 0|m_u\bar{u}u + m_d\bar{d}d)|0\rangle$$

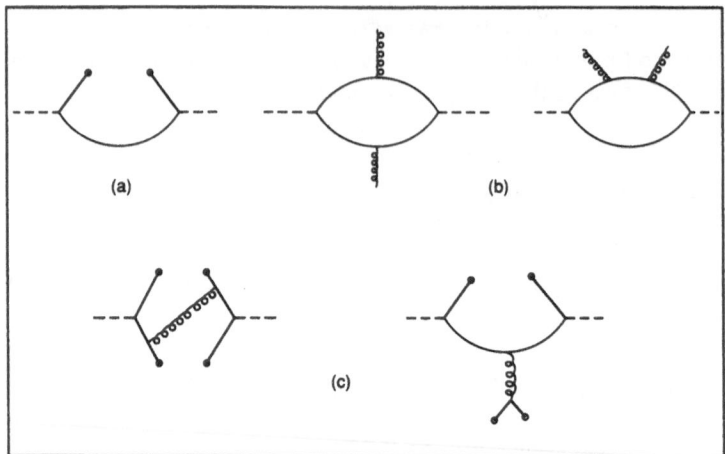

Fig. 9.2. Power corrections to Π. The diagram (a) is the quark condensate contribution. Diagrams (b) are the gluon condensate contribution and the diagrams (c) are the contributions from the four quark operators. In the vector channel they are the second, third and the last two terms respectively

$$+ \frac{1}{24Q^4} \langle 0 | \frac{\alpha_s}{\pi} G^a_{\mu\nu} G^a_{\mu\nu} | 0 \rangle - \frac{112\pi\alpha_s}{81Q^6} \langle 0 | \bar{q}q | 0 \rangle^2 \qquad (9.3.3)$$

To relate (2) to the phenomenological polarization of (1.3) we need to go low Q^2 region. This is because the light quarks and gluons within the hadron travel a large distance corresponding to a Q^2 beyond the asymptotic freedom limit where free propagation gets affected by the non-perturbative effects.

In order to probe large distances the prescription is to take derivatives with respect to Q^2. If Q^2 tends to infinity, the number of derivatives which can be reliably calculated is also large and one can consider the limit $Q^2 \Rightarrow \infty$, $n \Rightarrow \infty$, $Q^2/n \equiv M^2$ fixed. In effect a new parameter is introduced instead of Q^2. It corresponds to the Borel transform of $\Pi^j(Q^2)$:

$$L_M \Pi^j(Q^2) = \underset{\substack{Q^2, n \Rightarrow \infty \\ Q^2/n \equiv M^2}}{\text{Limit}} \frac{1}{(n-1)!} (Q^2)^n \left(-\frac{d}{dQ^2} \right)^n \Pi^j(Q^2) \qquad (9.3.4)$$

When applied to the theoretical side of the SR (2.2 or 3.1) an operator of dimension d is suppressed by a factor $1/(d/2 - 1)!$ On the phenomenological side the Borel transform of Eq. (1.3) becomes:

$$L_M \Pi^j(Q^2) = \frac{1}{\pi M^2} \int ds \, e^{-s/M^2} \, \text{Im} \, \Pi^j(s) \qquad (9.3.5)$$

The subtraction terms are eliminated and only a few resonances around M^2 need to be considered. The Borel transform should be flat and independent of M^2 in that region.

Taking Borel transform of (3), Eq. (5) can be written down as:

$$\frac{1}{\pi M^2} \int ds \ e^{-s/M^2} \ \text{Im} \ \Pi^j(s) = \frac{1}{8\pi^2} \left[1 + \frac{\alpha_s}{\pi} + \frac{8\pi^2}{M^4} \langle 0|m\bar{q}q|0\rangle \right.$$

$$\left. + \frac{\pi^2}{3M^4} \langle 0|\frac{\alpha_s}{\pi} G_{\mu\nu}^a G_{\mu\nu}^a|0\rangle - \frac{448\pi\alpha_s}{81M^6} \langle 0|\bar{q}q|0\rangle^2 \right] \tag{9.3.6}$$

From the charmonium sector the magnitude of the gluon condensate was fixed as $\langle 0|\alpha_s/\pi \, G_{\mu\nu}^a G_{\mu\nu}^a|0\rangle \cong (340 \ \text{MeV})^4$ and from current algebra one knows that

$$\langle 0|\bar{q}q|0\rangle = \frac{1}{2} f_\pi^2 K, \ K \equiv \frac{m_\pi^2}{m_u + m_d} \tag{9.3.7}$$

In the chiral limit m_u, $m_d \Rightarrow 0$, $m_\pi^2 \Rightarrow 0$, the quantity K remains finite. For $m_u + m_d \cong 15$ MeV, $\langle 0|\bar{q}q|0\rangle \cong -(250 \ \text{MeV})^3$ and $\langle 0|m\bar{q}q|0\rangle \cong -(100 \ \text{MeV})^4$. Even though there is a factor 1/24 in the gluon condensate term, the quark mass makes the quark condensate contribution five times smaller. But four quark operators contribute substantially.

On the phenomenological side one writes a narrow resonance and a continuum for spectral density, Im $\Pi(s)$:

$$\text{Im} \ \Pi(s) = \frac{\pi m_\rho^2}{g_\rho^2} \delta(s - m_\rho^2) + \frac{1}{8\pi} \left(1 + \frac{\alpha_s}{\pi} \right) \theta(s - s_0) \tag{9.3.8}$$

On taking the Borel transform one gets both sides of the sum rule. One can analyse the sum rule for power corrections, threshold dependence over a range of M^2. All these are given in details in the literature we have referred to. Finally, the parameters obtained are:

$$\frac{g_\rho^2}{4\pi} \cong 2.42, \quad m_\rho \cong 770 \ \text{MeV},$$

which compares very well with experimental values 2.36 ± 0.18 and 776 ± 3 MeV. The Borel parameter $M^2 \cong m_\rho^2$.

These results are not accidental. In fact, similar calculations were carried out for axial vector current $j_\mu = \bar{q}\gamma_\mu\gamma_5 q$. The correlator contains two independent functions Π_1 and Π_2:

$$\Pi_{\mu\nu}^A(Q^2) = i \int d^4x e^{iqx} \langle 0|T[j_\mu(x)\bar{j}_\nu(0)]|0\rangle$$

$$= -\Pi_1(Q^2)g_{\mu\nu} + \Pi_2(Q^2)q_\mu q_\nu$$

$$= q_\mu q_\nu (\Pi_2 - \Pi_1/q^2) + (q_\mu q_\nu - q^2 g_{\mu\nu})\Pi_1/q^2 \tag{9.3.9}$$

since the axial vector is not conserved in the real world. The spectral densities Im $\Pi_{1,2}$ contain not only the axial A_1 meson but also the pion.

$$\text{Im} \ \Pi_1 = \frac{\pi m_A^4}{g_A^2} \delta(s - m_A^2) + \frac{s}{8\pi} \left(1 + \frac{\alpha_s}{\pi} \right) \theta(s - s_0) \tag{9.3.10}$$

$$\text{Im}\,\Pi_2 = \frac{\pi}{2}f_\pi^2\delta(s) + \frac{\pi m_A^2}{g_A^2}\delta(s - m_A^2) + \frac{1}{8\pi}\left(1 + \frac{\alpha_s}{\pi}\right)\theta(s - s_0)$$

$\Pi_1 + Q^2\Pi_2$ corresponds to the longitudinal component which is related to the pion pole term leading to (7). Thus PCAC is rederived. In effect, two sum rules are derived, one through $\text{Im}\,\Pi_1$ and the other from $\text{Im}\,\Pi_2$. Consistency between the two led SVZ to the values $m_A = 1.15 \pm .04$ GeV and $4\pi/f_A^2 \cong 0.15$ to 0.18. Again the power corrections are found to be very important, specially the four quark operator and the gluon condensate. But in the pseudoscalar channels these are not enough.

Consider for example the π^0 current $j = i/\sqrt{2}(u\gamma_5 u - \bar{d}\gamma_5 d)$. The relevant correlator is

$$\Pi(Q^2) = \frac{3}{8\pi^2}Q^2\ln\frac{Q^2}{\mu^2} - \frac{1}{2Q^2}\langle 0|m_u\bar{u}u + m_d\bar{d}d|0\rangle + \frac{1}{8Q^2}\langle 0|\frac{\alpha_s}{\pi}G_{\mu\nu}^a G_{\mu\nu}^a|0\rangle$$

$$+ \frac{1}{Q^4}\langle 0|\mathcal{O}_1 + \mathcal{O}_2|0\rangle + \dots \tag{9.3.11}$$

where the four fermion operators are as follows:

$$\mathcal{O}_1 = -\pi\alpha_s(\bar{u}\sigma_{\mu\nu}\lambda^a u)(\bar{d}\sigma_{\mu\nu}\lambda^a d)$$

$$\mathcal{O}_2 = \frac{\pi\alpha_s}{2}[(\bar{u}\sigma_{\mu\nu}\lambda^a u)^2 + (\bar{d}\sigma_{\mu\nu}\lambda^a d)^2] + \frac{\pi\alpha_s}{3}(\bar{u}\gamma_\mu\lambda_u^a$$

$$+ \bar{d}\gamma_\mu\lambda^a d)\sum_{u,d,s}\bar{q}\gamma_\mu\lambda^a q$$

Recall (denoting $M^2 = 1/\tau^2$) $\Pi(\tau) = 1/\pi\int\text{Im}\,\Pi(s)e^{-s\tau^2}ds$. It could be separated into a free quark correlator (perturbative) and a non-perturbative correction term:

$$\frac{1}{\pi}\int_0^\infty \text{Im}\,\Pi^{per}(s)e^{-s\tau^2}ds + \Pi^{non\text{-}pert}(\tau)$$

where at large s the spectral density approaches to a *perturbative* result.

On the phenomenological side the spectral density consists of poles and a gap (s_0) after which the asymptotic regime starts. In this channel s_0 is rather large compared to the low-lying masses (first excited state of π is at 1300 MeV, $s_0 \cong (2$ GeV^2). It is approximated to $\text{Im}\,\Pi(s) = \pi\lambda^2\delta(s - m^2) + \theta(s_0 - s)\,\text{Im}\,\Pi^{pert}(s)$ which after Borelizing becomes

$$\frac{1}{\pi}\int_0^{s_0} \text{Im}\,\Pi^{per}(s)e^{-s\tau}ds + \Pi^{non\text{-}pert}(\tau) = \lambda^2 e^{-m^2\tau} \tag{9.3.12}$$

The $s > s_0$ part is essentially identical on both sides so that they cancel.

Since the pion is almost massless the above relation is independent of τ. In the limit $\tau \Rightarrow 0$ it gives the mass and the coupling, m_π and λ_π, $\lambda_\pi = \langle 0|j|\pi\rangle = f_\pi K$

where K is given by (7). With the same estimates of the condensates it was found out that the power corrections are too small to produce necessary s_0 so that the pion coupling to the pseudoscalar current is rather small. The validity of OPE in terms of non-perturbative operators is by no means obvious. SVZ argued that non-vanishing gluon condensate is due to large size instantons whereas the small size instantons break down the OPE itself. In fact *Novikov* et al. (1981) showed that all the spin-zero channels are the subject of strong non-perturbative effects. Shuryak made valiant attempts to include small size instantons which are of course one step forward to the power corrections. We refer the readers to his book (1988) for details.

The problem is rather difficult: the spontaneous breakdown of the chiral symmetry. In the pseudoscalar channels the strange and non-strange quarks are mixed so strongly that they produce almost ideal $SU(3)$ octet and singlet states (π, K, η, within 500 MeV and η' 958 MeV). Simple power corrections are insufficient to reproduce their masses and couplings.

As discussed in the Chapters 5 and 6, light quarks lessen the instanton density and the tunnelling amplitude between different θ-vacuum is negligibly small. But the chiral symmetry is already broken in QCD-vacuum. *Horn* and *Yankielovicz* (1978) showed the importance of instantons in evaluating the pion mass in the frame work of bag model. The question is then can one evaluate $\langle |\bar{q}q| \rangle$? *Caldi*, (1977) considered this in the mean field approach. Assuming $\langle |\bar{q}q| \rangle \neq 0$ one can ask further: what is the contribution of small instantons? Indeed, assymmetry of the vacuum allows to flip chirality of quarks emitted from such instantons. The result was interpreted as a generation of the effective quark mass (SVZ, 1980) m_*:

$$m_* = -\frac{2\pi^2}{3}\rho_c^2 \langle 0|\bar{\psi}\psi|0\rangle \tag{9.3.13}$$

In terms of instanton density n_c (when all of them have a size ρ_c):

$$\langle 0|\bar{\psi}\psi|0\rangle = -\frac{\left(\frac{3}{2}n_c\right)^{1/2}}{\pi\rho_c}, \text{ so that } m_* \cong 2\pi\rho_c(n_c/6)^{1/2} \tag{9.3.14}$$

The gluon condensate sets the upper limit for n_c. *Shuryak*, 1982 showed that for phenomenological success one needs a $\rho_c \cong 0.3 - 0.4$ fm and $n_c \cong 1$ fm^{-4}. The quark propagator is given by

$$S(x, y) = \frac{\psi_0(x)\psi_0^+(y)}{-m_*} + \dots \tag{9.3.15}$$

ψ_0 is the zero mode solution in the instanton background field (Eq. (6.7.2)). The corresponding contribution to the polarization operator can be calculated. Sum rules are found to be improved and a current mass of 150 MeV for the strange quark can explain the $SU(3)$ octet and η' mass splitting (*Shuryak*, 1983).

9.4   Introduction to Sum Rules for Baryons

The baryon currents are quite different from meson currents in that they have three quark wavefunctions. The right quantum numbers of spin and parity properties put strong constraints on the form of the current, but still one may choose more than one form for a particular baryonic channel, say the nucleon. The choice of a particular current may depend on one's intuition: so for example the original choice by *Ioffe* (1981), which we give below, contains the desirable features of having the least contribution from perturbative terms and higher resonances:

$$\eta_N = \varepsilon_{abc} \left(u_a^T(x) C \gamma_\mu u_b(x) \right) \gamma_5 \gamma_\mu d_c(x) \tag{9.4.1}$$

Here T indicates transposed and C is the charge conjugation operator. ε_{abc} makes the current colour singlet. The currents can also be written in different equivalent forms, the above current is the special case of the combination $A = -B = 1$ of the general combination

$$\eta_N = \varepsilon_{abc} \left\{ A \left(u_a^T(x) C d_b(x) \right) \gamma_5 + B \left(u_a^T(x) C \gamma_5 d_b(x) \right) \right\} u_c(x) \tag{9.4.2}$$

There are more choices, but they have to tackle the baryons as well as baryon resonances together (*Y. Chung* et al. 1981, 1982). Even if one excludes currents which could have the operator ∂_μ there is a third form

$$\eta_N = \varepsilon_{abc} \left(u_a^T(x) C \sigma_{\mu\nu} u_b(x) \right) \gamma_5 \sigma_{\mu\nu} d_c(x) \tag{9.4.3}$$

but it was shown by *Espriu* et al. (1983), that this contributes only 2% to the final results. In short, whatever the current one chooses the results for the nucleon mass come out to be the same and generally there is a trend to use the current given in Eq. (1). Masses of the octets and decuplets were also found out by *Belyaev* and *Ioffe* (1983) fixing the ratio of the strange quark condensate to be 17% less than that of the light quark condensate in these baryons.

9.5   The Nucleon Correlator

The polarization operator for the current η_N has a Dirac structure:

$$i \int d^4x \, e^{iqx} \langle 0 | T(\eta_N(x) \eta_N(0)) | 0 \rangle = \Pi_s(q^2) + q_\mu \gamma_\mu \Pi_q(q^2) \tag{9.5.1}$$

This is also reflected on the phenomenological side, which we write in terms of the nucleon pole (propagator) and the continuum

$$\Pi(q^2) = -\lambda_N^2 \frac{q_\mu \gamma_\mu + M_N}{q^2 - M_N^2} + \text{continuum.} \tag{9.5.2}$$

λ_N^2 is the coupling of the nucleon to the generic nucleon current which actually carries all the information in the channel. The value of this coupling therefore will depend on the choice of the current, for example on A and B in Eq. (4.2).

In Eq. (1) since Π_s has no q_μ, only the odd dimensional operators will contribute to it, the leading term being the $\langle \bar{q}q \rangle$ condensate (without a factor m_q in front) whereas to Π_q only the even dimension, the identity (I) and $\langle G_{\mu\nu}^2 \rangle$ contribute with the latter only 5% of the leading term! Thus the gluons play a relatively minor role in the structure of the nucleon, a fact which is easily acceptable to nuclear physicists. The other part, $\langle \bar{q}q \rangle$, controls the mass as we shall see soon. This is quite surprising since one imagines the diquark to be like the quark, and then one would expect that like in the ρ-meson both the gluon and the quark condensates should contribute to the mass. Similar Regge trajectories support this picture. It is sometimes argued that the quark condensate changes fast with density or temperature whereas the gluon condensate remains more or less constant. This might imply different scaling laws for mesons and baryons.

9.6 Details of the Sum Rule for Nucleons

Let us illustrate the calculation with some details, since the nucleon is the heart of nuclear physics, and this is the only chance to see clearly how that heart is constructed! The lattice calculations are never so clear analytically. The uninterested reader may skip this section jump to Eq. (7). We have

$$i \int d^4x e^{iqx} T[\eta_N(x)\eta_N(0)] = C_d(q^2)\bar{d}d + C_u(q^2)\bar{u}u \qquad (9.6.1)$$

Let us pick the first term on the right hand side which arises when we contract the u-quarks in the two currents but leave out the d-s (in colour states c, c'):

$$C_d(q^2)\bar{d}_c(p)d_{c,}(p)\delta_{cc'} = i \int d^4x e^{iqx} \langle p, c|T[\eta_N(x)\eta_N(0)]|p, c'\rangle$$

$$= -i \int d^4x e^{i(q-p)x} \varepsilon_{abc}\varepsilon_{a'b'c'}\gamma_5\gamma_\mu d_c(p)\bar{d}_{c'}(p)\gamma_5\gamma_\nu$$

$$\times \langle 0|u_a^T(x)C\gamma_\mu u_b(x)\bar{u}_{b'},(0)\gamma_\nu Cu_{a'}^T,(0)|0\rangle \qquad (9.6.2)$$

where we have used the fact that $C^T = -C = C^*$.

There are two possible contractions for the $u-s$. The trace can be evaluated using standard formulae. The factor $\bar{d}(p)d(p) = 1$, but $d(p)\bar{d}(p) = (p+m)/(4m) = 1/4$ in the limit $p \Rightarrow 0$. There is an integration over the loop momentum which is divergent and thus tricky. A cutoff Λ is used:

$$C_d(q^2) = -\frac{1}{(2\pi)^2}q^2 \ln\left(\frac{\Lambda^2}{-q^2}\right), \quad C_u(q^2) = 0$$

$$\Pi_s(q^2) = \frac{1}{4\pi^2}\langle 0|\bar{q}q|0\rangle q^2 \ln\left(\frac{-q^2}{\Lambda^2}\right) \qquad (9.6.3)$$

$$\Pi_q(q^2) = -\frac{1}{4(2\pi)^4} q^4 \ln\left(\frac{-q^2}{\Lambda^2}\right)$$

$$-\frac{1}{2(4\pi)^2} \langle 0|\frac{\alpha_s}{\pi} G^a_{\mu\nu} G^a_{\mu\nu}|0\rangle \ln\left(\frac{-q^2}{\Lambda^2}\right) - \frac{2}{3q^2} \langle 0|\bar{q}q|0\rangle^2 + \dots \quad (9.6.4)$$

Compare Eqs. (5.1) and (5.2). On the phenomenological side we have M_N and $q_\mu\gamma_\mu$ terms. They correspond to two invariant functions Π_s and Π_q respectively. Taking Borel transform on both sides we arrive at two sum rules, one from Π_s:

$$2aM^4 = 2(2\pi)^4 M_N \lambda_N^2 \exp(-M_N^2/M^2) \quad (9.6.5)$$

and another from Π_q:

$$M^6 + bM^2 + (4/3)a^2 = 2(2\pi)^4 \lambda_N^2 \exp(-M_N^2/M^2) \quad (9.6.6)$$

where $a = -(2\pi)^2 \langle 0|\bar{q}q|0\rangle$ and $b = \pi^2 \langle 0|\alpha_s/\pi G^a_{\mu\nu} G^a_{\mu\nu}|0\rangle$.
From the above Eqs. (5) and (6) we see that in the leading order one gets the nucleon mass (setting the Borel mass scale $M \cong M_N$)

$$M_N \cong \{2(2\pi)^2 \langle 0|\bar{q}q|0\rangle\}^{1/3} \cong 1 \text{ GeV} \quad (9.6.7)$$

Results of *Ioffe*, (1981) show that the sum rules for Π_s and Π_q are rather sensitive to the variation of M^2. Power corrections of higher order, difference of thresholds (*Belyaev* and *Ioffe*, 1983) and next-to-leading order corrections in α_s of perturbative QCD theory (*Ovchinnikov* et al. 1988; *Jamin*, 1988) have been added to improve stability. A careful analysis (*Henley* and *Pashupathy*, 1992) shows that all these make Π_q stable but not Π_s. Again as in pseudoscalar channels, *Dorokhov* and *Kochelev*, (1990) found that small size instantons stabilizes the Π_s sum rule remarkably well. The OPE is not enough when the leading term itself is a non-perturbative one.

9.7   Finite Temperature and Density

Much work has been done to adapt the QCD sum rule technique to study the hadron properties at finite temperature and density (*Dey, Dey* and *Ghose*, 1985; *Bochkarev* and *Shaposnikov*, 1986; *Bochkarev*, 1987; *Dominguez* and *Loewe*, 1989, 1991; *H. G. Dosch* and *S. Narison*, 1988; *Adami, Hatsuda* and *Zahed*, 1991; *Furnstahl, Hatsuda* and *Lee*, 1990). Take the case of finite temperature. In the presence of a heat bath one has to take the thermal average of an operator $\hat{O}$:

$$\langle \hat{O} \rangle_T = \frac{\sum_n \langle n|e^{-\beta H} \hat{O}|n\rangle}{\sum_n \langle n|e^{-\beta H}|n\rangle} \quad (9.7.1)$$

where H is the QCD Hamiltonian and the sum is over all states. The above averaging has to be performed for *each operator* in the OPE. Wilson coefficients are evaluated perturbatively at deep Euclidean region. But the thermal part of the quark propagator is on the mass shell. In the light quark sector this part is well overlapping with the nonperturbative sector. Hence to avoid the short distance and long distance mixture one emblems the operators T dependent whereas Wilson coefficients remain unheated:

$$\Pi(q, T) = \sum C_n(q)\langle \hat{O} \rangle_T \tag{9.7.2}$$

Again, the presence of $e^{-\beta H}$ in (1) breaks the Lorentz invariance. In addition to scalar operators, listed in (2.2), non-scalar operators like $\bar{q}\gamma_\mu D_\nu q$, $\bar{q}\gamma_\mu D_\nu D_\lambda D_\rho q$ etc. will appear in the OPE of the correlator. In other words,

$$\Pi(q, t) = \Pi_0(q, T) + \Pi_{NS}(q, T) \tag{9.7.3}$$

At low temperature $T \sim F_\pi$, these non-scalar operators are of order T^4 (*Hatsuda* et al. 1992) whereas the scalar operators are of order T^2. Even in this order *Dey*, *Eletsky* and *Ioffe*, (1990) showed that the vector and axial-vector channels could be mixed when the sum (1) is saturated with soft thermal pions, signalling towards the restoration of chiral symmetry. At high T the situation is very different. *Eletsky* and *Ioffe* (1988) showed that a meson correlator goes as $2\pi T$. The quarks behave as if they are free with a "thermal" mass πT. This behaviour has since been observed in lattice calculations as discussed in Chapter 1.

Much effort has also been put in to develop QCD sum rules at finite density. This is highly interesting from the point of view of nuclear physics. Basically, instead of 2 invariant functions Π_s and Π_q we have Π_p as well. This is because of the momentum **p** of the nucleon in the medium (*Drukarev* and *Levin*, 1990). From there one gets the scalar as well as vector potential (*Cohen* et al. 1991). *Henley* and *Pashupathy*, (1992) showed that the sum rule obtained from Π_s remains problematic even in the medium. As shown by Dorokov and Kochelev at zero density instantons may have a stabilizing influence here too. These studies are far from complete and it is time now to put a lot more effort to understand nuclear physics from QCD.

References

Accetta, F., Caldi, D. G., and Chodos, A.: Phys. Lett. **B226** 175 (1989)

Actor, A.: Rev. Mod. Phys. **51**, 461 (1979)

Adami, C., Hatsuda, T., and Zahed, I.: Phys. Rev. **D43** 921 (1991)

Adkins, G., Nappi, C., and Witten, E.: Nucl. Phys. **B228** 552 (1983)

Adler, S.: Phys. Rev. **177** 2426 (1969)

Aerts, A., and Rafelski, J.: Phys. Lett. **B148** 337 (1984)

Affleck, I.: Nucl. Phys. **B191** 445 (1981)

Altarev, I. S., et al.: JETP Lett. **44** 460 (1986)

Aliev, T. M., and Shifman, M. A.: Sov. J. Nucl. Phys. **36** 891 (1982)

Ansari, A., Dey, J., Dey, M., Matin, M. A., and Ghose, P.: Hadronic Journal Supplement **5** 233 (1990)

Anselmino, M., and Scadron, M. D.: Phys. Lett. **B229** 117 (1989)

Applequist, T., and Carrazone, J.: Phys. Rev. **D11** 2856 (1975)

Atyiah, M., Drienfeld, V., Hitchin, N., and Manin, Y.: Phys. Lett. **A65** 185 (1978)

Atiyah, M., and Manton, N.: Phys. Lett. **B222** 438 (1989) and Oxford preprints (to be published) 1992

Atiyah, M., and Singer, I.: Ann. Math. **87** 484 (1968)

Atiyah, M., and Ward, A.: Comm. Math. Phys. **55** 117 (1977)

Balachandran, A., Nair, V. P., Rajeev, S., and Stern, A.: Phys. Rev. Lett. **49** 1124 (1982); **E50** 1630 (1983); Phys. Rev. **D27** 1153 (1983)

Bando, M., Kugo, T., and Yamazaki, K.: Phys. Rep. **C164** 217 (1988) and references therein

Barut, A. O.: Phys. Rev. **B135** 839 (1964); Phys. Lett. **B26** 308 (1968)

Belavin, A., Polyakov, A., Schwartz, A., and Tyupkin, Y.: Phys. Lett. **B59** 85 (1975)

Bell, J., and Jackiw, R.: Nuovo Cim. **60A** 47 (1969)

Belyaev, V., and Ioffe, B.: JETP **57** 716 (1983)

Bernard, C., Phys. Rev. **D19** 3013 (1979)

Bhaduri, R. K.: *Models of the Nucleon* (Addison Wesley, 1988)

Bhaduri, R. K., and Brack, M.: Phys. Rev. **D25** 1443 (1982)

Bhaduri, R. K., Cohler, J., and Nogami, Y.: Phys. Rev. Lett. **44** 1369 (1980); Nuovo Cim. **65A** 376 (1981)

Bhaduri, R. K., and Dey, M.: Phys. Lett. **B125** 513 (1983)

Bhaduri, R. K., Dey, J., and Preston, M. A.: Phys. Lett. **B136** (1984)

Bhaduri, R. K., Dey, J., and Srivastava, M. K.: Phys. Rev. **D31** 1765 (1985)

Bhaduri, R. K., Suzuki, A., Abdalla, H., and Preston, M. A.: Phys. Rev. **D41** 959 (1990)

Bienkowska, J., Dziembowski, Z., and Weber, H.: Phys. Rev. Lett. **59** 624 (1987)

Birse, M. C.: Prog. Part. Nucl. Phys. **25** 1 (1990); Chiral Solitons, preprint Manchester TH 92/03, Invited talk at Conference on Medium and High Energy Nuclear Physics, Calcutta 20–22 December, 1991

Birulev, V. K., et al.: Nucl. Phys. **B182** 1 (1981)

Bjorken, J., and Drell, S.: *Relativistic Quantum Mechanics* (McGraw Hill, 1964); *Relativistic Quantum Fields* (McGraw Hill, 1965)

Bochkarev, A. I., and Shaposhnikov, M. E.: Nucl. Phys. **B268** 220 (1986)

Bochkarev, A. I.: Sov. J. Nucl. Phys. **45** 517 (1987)

Born, K., et al.: (MT$_c$ collaboration), Phys. Rev. Lett. **67** 302 (1991)

Bourdeau, M., and Mukopadhyay, N.: Phys. Rev. Lett. **58** 976 (1987)

Boyer, T.: Ann. Phys. **59** 474 (1970)

Brevik, I.: Phys. Rev. **D33** 290 (1986)

Brown, F., et al.: Phys. Rev. Lett. **67** 1062 (1991)

Brown, G. E., Kubodera, K., Prakash, M., and Rho, M.: Nucl. Phys. **A479** 175c (1988)

Brown, G. E., and Rho, M.: Phys. Lett. **B222** 324 (1989); Phys. Lett. **B237** 3 (1990); Phys. Rev. Lett. **66** 2720 (1991)

Buchmüller, W., and Wyler, D.: Phys. Lett. **B121** 321 (1983)

Caldi, D. G.: Phys. Rev. Lett. **39** 121 (1977)

Caldi, D., and Chodes, A.: Phys. Rev. **D36** 2876 (1987)

Callan, C., Dashen, R., and Gross, D.: Phys. Rev. **D17** 2717 (1978)

Carlitz, R., and Creamer, D.: Ann. Phys. **116** 429 (1979)

Carlson, J.: Nucl. Phys. **B253** 149 (1985)

Capstick, S., and Isgur, N.: Phys. Rev. **D34** 2809 (1986)

Cenni, R., Conte, F., and Dillon, G.: Lett. Nuovo Cim. **43** 39 (1985)

Cho, Y., et al.: Phys. Rev. **D22** 2688 (1980)

Chodos, A., Jaffe, R., Johnson, K., Thorn, C., and Weisskopf, V.: Phys. Rev. **D9** 3471 (1974)

Chung, Y., Dosch, H., and Kremer, M., and Schall, D.: Phys. Lett. **B102** 175 (1981); Nucl. Phys. **B197** 55 (1983)

Close, F.: *An Introduction to Quarks and Partons* (Academic Press, London, 1979)

Close, F., Roberts, R., and Ross, G.: Phys. Lett. **B168** 400 (1986)

Coleman, S., and Witten, S.: Phys. Rev. Lett. **45** 100 (1980)

Crater, H. W., and Van Alstine, P.: Phys. Rev. Lett. **53** 1527 (1984)

Cudell, J., and Dienes, K.: Phys. Rev. Lett. **69** 1324 (1992)

Cohen, T., Furnstahl, R., and Griegel, D.: Phys. Rev. Lett. **67** 961 (1991)

De Tar, C., and Kogut, J.: Phys. Rev. Lett. **59** 399 (1987)

Dey, J., and Dey, M.: Phys. Lett. **B138** 200 (1984); **176** 469 (1986); **234** 349 (1990)

Dey, J., Dey, M., and Le Tourneux, J.: Phys. Rev. **D34** 2104 (1986)

Dey, J., Dey, M., Ghose, P., and Mukhopadhyay, G.: Phys. Lett. **B209** 330 (1988)

Dey, J., Dey, M., and Ghose, P.: Phys. Lett. **B165** 181 (1985); **193** 98 (1987); **221** 161 (1989)

Dey, J., Tomio, L., and Dey, M.: Mod. Phys. Lett. **A5** 1451 (1990)

Dey, J., Dey, M., Mukhopadhyay, G., and Samanta, B. C.: Can. J. Phys. **69** 749 (1991a)

Dey, J., Dey, M., and Volkovitsky, P.: Phys. Lett. **B261** 493 (1991)

Dey, J., *Pramana* **37** 57 (1991)

Dey, J., Tomio, L., Dey, M., and Frederico, T.: Phys. Rev. **C44** 2181 (1991b)

Dey, J., Dey, M., and Tomio, L.: Phys. Lett. **B288** 306 (1992)

Dey, M., Eletsky, V., and Ioffe, I.: Phys. Lett. **B252** 620 (1990)

Dittrich, W., and Reuter, M.: *Selected Topics in Gauge Theories* (Lecture notes in Physics 244, Springer-Verlag, 1985)

Donaldson, G., et al.: Phys. Rev. **D9** 2960 (1974)

Donoghue, J., Holstein, B., and Klimpt, S.: Phys. Rev. **D35** 934 (1987)

Dorokhov, A. E., and Kochelev, N. I.: Sov. J. Nucl. Phys. **52** 135 (1990a); Mod. Phys. Lett. **A5** 55 (1990b); Z. Phys. **C45** 281 (1990c)

Domingüez, C. A., and Loewe, M.: Phys. Lett. **B233** 201 (1989); Z. Phys. **C51** 69 (1991)

Dosch, H. G., and Narison, S.: Phys. Lett. **B203** 155 (1988)

Dreiner, H., Ellis, J., and Flores, R.: Phys. Lett. **B221** 166 (1989)

Drukarev, E. G., and Levin, E. M.: Nucl. Phys. **A511** 679 (1990), **A516** 715(E) (1990)

Dynakov, D. I., and Petrov, V. Y.: Nucl. Phys. **B245** 259 (1984)

Eletsky, V., and Ioffe, B.: Sov. J. Nucl. Phys. **48** 602 (1988)

Engels, J., et al.: Phys. Lett. **B252** 625 (1990)

Espriu, D., Pascual, D., and Terrach, R.: Nucl. Phys. **B214** 285 (1983)

Fai, G., Perry, R., and Wilets, L.: Phys. Lett. **B208** 1 (1988)

Fiolhais, M., Neuber, T., Goeke, K., Alberto, P., and Urbano, J. N.: Phys. Lett. **B268** 1 (1991)

Fleck, S., and Richard, J.: Part. World **1** 67 (1990)

Forte, S.: Phys. Lett. **B224** 189 (1989); Nucl. Phys. **B331** 1 (1990)

Franklin, J., Lichtenberg, D., Namgung, W., and Carydas, D.: Phys. Rev. **D24** 2910 (1981)

Frederico, T., Carlson, B., Rego, R., and Hussein, M.: J. Phys. **G15** 297 (1989)

Freund, P., and Rosner, J.: Phys. Rev. Lett. **68** 765 (1992)

Friedberg, T., and Lee, T. D.: Phys. Rev. **D15** 1694 (1977); **D16** 1096 (1977); **D18** 2623 (1978)

Fritzsch, H.: Mod. Phys. Lett. **A5** 1815 (1990) (brief review)

Fujikawa, K.: Phys. Rev. Lett. **42** 1195 (1979); Phys. Rev. **D21** 2848 (1980)

Fukugita, M., Okawa, M., and Ukawa, A.: Phys. Rev. Lett. **63** 1768 (1989); Nucl. Phys. **B337** 181 (1990)

Furnstahl, R., Hatsuda, T., and Lee, S.: Phys. Rev. **D42** 1744 (1990)

Gasser, J., and Leutwyler, H.: Phys. Rep. **87C** 77 (1982); Ann. Phys. **158** 142 (1984); Nucl. Phys. **B250** 485 (1985)

Gasser, J., Leutwyler, H., and Sainio, M.: Phys. Lett. **B253** 252, 260 (1991)

Gasser, J., Sainio, M. E., and Svarc, A.: Nucl. Phys. **B307** 779 (1988)

Gavai, R., Potvin, J., and Sanielevici, S.: Phys. Rev. **D40** 2743 (1989)

Gavai, R., et al.: (MT$_c$ collaboration), Phys. Lett. **B241** 567 (1990)

Gerber, P., and Leutwyler, H.: Nucl. Phys. **B321** 387 (1989)

Georgi, H.: Phys. Rev. Lett. **63** 1917 (1989)

Gershtein, S., and Dzhikiya, D.: Sov. J. Nucl. Phys. **31** 870 (1981)

Glashow, S., Iliopoulos, J., and Maiani, L.: Phys. Rev. **D2** 1285 (1970)

Gottlieb, S., et al.: Phys. Rev. Lett. **59** 1515, 1881 (1987a); 2247 (1987b); Phys. Rev. **D51** 662 (1990)

Gocksch, A., Rossi, P., and Heller, U.: Phys. Lett. **B205** 3345 (1988)

Gross, D., and Wilczek, F.: Phys. Rev. Lett. **26** 1343 (1973)

Guichon, P. A. M.: Phys. Lett. **B200** 235 (1988)

Gupta, S.: Phys. Lett. **B288** 171 (1992)

Hatsuda, T., Koike, Y., and Lee, S.: Univ. of Maryland PP # 92–203 Nucl. Phys. B (in press)

Hagedorn, R., and Rafelski, J.: Phys. Lett. **B97** 180 (1980); in *Statistical Mechanics of Quarks and Gluons*, ed. Satz. H., (N. Holland, 1981)

Harrington, B., and Shepard, H.: Phys. Rev. **D17** 2122 (1978)

Henley, E., and Pashupathy, J.: University of Washington Preprint, July, 1992

Hill, D., et al.: Nucl. Phys. **B153** 39 (1979)

Horn, D., and Yankielowicz, S.: Phys. Lett. **B76** 343 (1978)

Huang, K., and Stump, D.: Phys. Rev. **D14** 223 (1976)

Ioffe, B.: Nucl. Phys. **B188** 175 (1981), Errata **B191** 591 (1981)

Ioffe, B., and Kerliner, M.: Phys. Lett. **B247** 387 (1990)

Isgur, N., and Karl, G.: Phys. Rev. **D18** 4187 (1978); **D19** 2653 (1979)

Isgur, N., Karl, G., and Koniuk, R.: Phys. Rev. **D25** 2394 (1982)

Itzykson, C., and Zuber, J.: *Quantum Field Theory* (McGraw Hill, 1980)

Jackiw, R., and Rebbi, C.: Phys. Rev. **D13** 3398 (1976)

Jackiw, R., Nohl, C., and Rebbi, C.: Phys. Rev. **D15** 1642 (1977)

Jackiw, R.: in *Anomalies, Geometry, Topology*, ed. Bardeen, W. A., and White, A. R., (World Scientific, Singapore, 1985); Comm. on Nucl. and Part. Phys. **13** 27 (1984)

Jackiw, R., and Rajaraman, R.: Phys. Rev. Lett. **54** 1219 (1985)

Jackson, A. D., and Rho, M.: Phys. Rev. Lett. **51** 751 (1983)

Jackson, A., Jackson, A. D., and Pasquier, V.: Nucl. Phys. **A438** 567 (1985)

Jaffe, R., and Lipkin, H.: Phys. Lett. **B266** 458 (1991)

Jain, P., Schechter, J., and Sorkin, R.: Phys. Rev. **D39** 998 (1989)

Jamin, M.: Z. Phys. **C37** 635 (1988)

Jennings, B. K., and Bhaduri, R. K.: Phys. Rev. **D26** 1750 (1982)

Kalman, C. S., and Tran, B.: Nuovo. Cim. **A102** 835 (1989)

Kapusta, J.: Phys. Rev. **D23** (1981) 2444; Nucl. Phys. **B196** 1 (1982)

Khatsimovskii, V. M., et al.: Zeits. Phys. **C36** 455 (1987)

Koch, P., Shuryak, E., Brown, G. E., and Jackson, A. D.: Phys. Rev. **D46** 3169 (1992)

Kogut, J., Sinclair, D., and Wang, K.: Phys. Lett. **B263** 101 (1991)

Klinkhamer, F., and Manton, N.: Phys. Rev. **D30** 2212 (1984)

Kochelev, N.: Sov. J. Nucl. Phys. **41** 291 (1985)

Koniuk, A., and Isgur, N.: Phys. Rev. **D21** 1868 (1980)

Krein, G., Tang, P., Wilets, L., and Williams, A.: Nucl. Phys. **A523** 548 (1991)

Kutasov, D., and Seiberg, N.: Nucl. Phys. **B358** 600 (1991)

Kuzmin, V., Rubakov, V., and Shaposhnikov, M.: Phys. Lett. **B155** 36 (1985)

Lacombe, M., Loiseau, B., Vinh Mau, R., and Cottingham, W. N.: Phys. Rev. **D38** 1491 (1988)

Leal Ferreira, P.: Lett. Nuovo Cim. **20** 157 (1977); Phys. Rev. **D38** 2648 (1989)

Leal Ferreira, P., and Zagury, N.: Lett. Nuovo Cim. **20** 511 (1977)

Leal Ferreira, P., Helayel, J., and Zagury, N.: Nuovo Cim. **55A** 215 (1980)

Lee, T. D.: *Particle Physics and Introduction to Field Theory* (Harwood Academic 1981)

Leech, R. G., and Birse, M. C.: J. Phys. **G18** 785 (1992)

Leite Lopes, J.: *Gauge Field Theories, An Introduction.* (Pergamon press, Oxford, 1981)

Leutwyler, H., and Roos, M.: Zeits. f. Phys. **C25** 91 (1984)

Lucha, W., Schöberl, F., and Gromes, D.: Phys. Rep. **C200** 127 (1990)

Matsui, T., and Satz, H.: Phys. Lett. **B178** 416 (1986)

McLerran, L.: Rev. Mod. Phys. **58** 1021 (1986)

McLerran, L., Vainshtein, A., and Voloshin, M.: Phys. Rev. **D42** 171 (1990)

Mignaco, J., and Wulck, S.: Phys. Rev. Lett. **62** 1449 (1989)

Milton, K.: Phys. Rev. **D27** 439 (1983)

Mitra, P., and Rajaraman, R.: Phys. Lett. **B225** 267 (1989)

Murthy, M., Dey, M., Dey, J., and Bhaduri, R. K.: Phys. Rev. **D30** 152 (1984)

Murthy, M., and Bhaduri, R. K.: Phys. Rev. Lett. **54** 745 (1985)

Muta, J.: *Foundations of Quantum Chromodynamics* (World Scientific, 1987)

Narison, S.: *QCD Spectral Sum Rules* (World Scientific, 1990)

Nagamiya, S.: Nucl. Phys. **A488** 3c–30c (1988). See 24–27c for charm production

Nielsen, H., and Ninomiya, M.: Phys. Lett. **B130** 389 (1983)

Nielsen, H., and Olesen, P.: Nucl. Phys. **B61** 45 (1973)

Nilles, H. - P.: Phys. Rep. **C110** 1 (1984)

Novikov, V., Shifman, M., Vainshtain, A., and Zakharov, V.: Nucl. Phys. **B191** 301 (1981)

Novozhilov, V. Y.: Phys. Lett. **B228** 240 (1989)

Nowak, M., and Zahed, I.: Phys. Lett. **B230** 108 (1989)

Oka, M., and Takeuchi, S.: Phys. Rev. Lett. **63** 1780 (1989); Nucl. Phys. **A524** 649 (1991)

Ovchinnikov, A., Pivovarov, A., Surguladze, L.: Yad. Fiz. **48** 562 (1988)

Palladino, B., and Leal Ferreira, P.: Phys. Rev. **D40** 3024 (1989)

Pascual, P., and Tarrach, R.: *QCD Renormalzation for the Practitioner* (Lecture Notes no. 194, Springer Verlag, 1984)

Peccei, R., and Quinn, H.: Phys. Rev. Lett. **38** 1440 (1977); Phys. Rev. **D16** 1781 (1977)

Pokorski, S.: *Gauge Field Theories*, (Cambridge Univ. Press, 1987)

Polchinski, J., and Wise, M.: Phys. Lett. **B125** 393 (1983)

Politzer, H. D.: Phys. Rev. Lett. **30** 1346 (1973)

Preparata, G., Ratcliffe, P., and Soffer, J.: Phys. Lett. **B273** 306 (1991)

Rabinowitz, S., et al.: Columbia preprint, 1992

Rajaraman, R.: *Solitons and Instantons* (North Holland, 1982)

Reinders, L., Rubinstein, H., and Yazaki, S.: Nucl. Phys. **B186** 109 (1981); **B196** 125 (1982)

Reinders, L., Rubinstein, H., and Yazaki, S.: Phys. Rep. **C127** 1 (1985)

Richardson, J.: Phys. Lett. **B82** 272 (1979)

Ringwald, A.: Nucl. Phys. **B330** 1 (1990)

de Rujula, A., Georgi, G., and Glashow, S.: Phys. Rev. **D12** 147 (1975)

Salpeter, E., and Bethe, H.: Phys. Rev. **84** 1232 (1951)

Satz, H.: Nucl. Phys. **A488** 511c (1988)

Schwinger, J., de Raad, L., and Milton, K.: Ann. Phys. **115** 1 (1979)

Semenoff, G. W., and Sodano, P.: Phys. Rev. Lett. **57** 1195 (1986)

Sen, A.: Comm. on Nucl. and Part. Phys. **A20** 23 (1991)

Shifman, M., Vainshtein, A., and Zakharov, V.: Nucl. Phys. **B147** 385, 448 (1979); **163** 46 (1980); **B165** 45 (1980)

Shuryak, E.: *QCD Vacuum, Hadrons and the Superdense Matter* (World Scientific, 1988)

Shuryak, E.: Nucl. Phys. **B203** 93 (1982); **214** 237 (1983); **302** 539, 574, 621 (1988)

Shuryak, E., and Rosner, J.: Phys. Lett. **B218** 72 (1989)

Shuryak, E., and Verbaarschot, J.: Nucl. Phys. **B341** 1 (1991)

Silvestre-Brac, B., and Gignoux, C.: Phys. Rev. **D32** 743 (1985)

Skyrme, T. H. R.: Proc. R. Soc. Lon. **A260** 127 (1961); Nucl. Phys. **31** 556 (1962)

Smith, K. F., et al.: Phys. Lett. **B234** 191 (1990)

Soni, V.: Phys. Lett. **B93** 101 (1980)

Su, W., Schrieffer, J., and Heeger, A.: Phys. Rev. Lett. **42** 1698 (1979); Phys. Rev. **B22** 2099 (1980)

Su, Z., and Sakita, B.: Phys. Rev. Lett. **56** 780 (1986)

Suzuki, A., Nogami, Y., and Tomio, L.: Prog. Th. Phys. **75** 880 (1986)

Takeuchi, S., and Oka, M.: Phys. Rev. Lett. **66** 1271 (1991)

Tegen, R., Brockmann, R., Weise, W.: Zeits. f. Phys. **A307** 339 (1982)

't Hooft, G.: Nucl. Phys. **B72** (1974) 461; Phys. Rev. Lett. **37** 8 (1976a); Phys. Rev. **D14** 3432 (1976b); Errata **D18** 2199 (1978)

Toki, H., Dey, J., and Dey, M.: Phys. Lett. **B133** 20 (1983)

Tomio, L., and Nogami, Y.: Phys. Rev. **D31** 2818 (1985)

Ukawa, A.: Nucl. Phys. B (Proc. Suppl.) **10A** 66 (1989)

Weinberg, S.: Phys. Rev. Lett. **37** 657 (1976); Physica **A96** 327 (1979)

Wilets, L.: *Nontopological Solitons* (World Scientific, 1989)

Wilson, K. G.: Phys. Rev. **179** 1499 (1969)

Witten, E.: Nucl. Phys. **B160** 57 (1979)

Zu-Rong Yu.: Mod. Phys. Lett. **A3** 1059 (1988)

Zahed, I., and Brown, G. E.: Phys. Rep. **C142** 1 (1986)

Note Added in Proof

The main text of our book was completed in December 1992 and we mention here a few interesting new developments. We start with lattice calculations, referred to at the end of chapter one.

The proceedings have now been published. We were excited by two interesting papers. *DeTar* [1] discussed the possibility of hadrons of many sizes in the hot phase of quark plasma. The situation is far from clear, but is very interesting. It is found in lattice calculations that quark gluon plasma (or even pure glue) in the high-T phase retains confinement of colour singlets, while bulk quantities like energy density are consistent with a nearly free gas of quarks. A resolution of this seeming paradox describes the quark gluon plasma as an ensemble of colour singlet clusters of various sizes. This result seems very attractive to us. Different sizes for different hadrons emerge from a model calculation using the Bekenstein entropy bound [2]. For details see [3].

The other paper is by Gupta where he discusses his earlier finding (*Gupta 1992*, we have already discussed) that there is strong residual interaction in the pion channel. Does this imply a pion pole? It was thought that since pion mass goes as πT, perhaps the pion survives as a fundamental field in the hot phase. *S. Gupta* [4] has shown that "fundamental" pions in the form of bosonic poles are not indeed present in the recent lattice simulations. A composite quark-antiquark, interacting strongly, can indeed have mass πT in a surprisingly simple model calculation [3].

The gold on gold heavy ion collision results have also appeared [5] and show that protons come out with a typical temperature T_{HI} in heavy ion collisions and pions with a lower one, unlike in proton–proton or in proton–nucleus collisions. For details and an attempted explanation, see [3]. In the experiment deuterons come out with T_{HI} even higher than that for protons, so it cannot be that protons coalesce into deuterons!

Other experimental results that are interesting concern deep inelastic scattering. 800 GeV protons on deuteron, carbon and tungsten have given data on the sea quark asymmetry of the u and d [6], found first in the Gottfried sum rule [7]. This can be explained in a model with an effective temperature [8]. Finally there is data from the NMC group which shows that the proton structure function does not dip at small x, but seems to continue to increase [9]. There is a long letter by *Dorokhov* and *Kochelev* [10] on the effect of instantons on spin-dependent proton structure function and the "bible" by *Close* [11] which seems to suggest that the spin crisis is not so serious. Finally there are excellent review articles by *Reya* [12] and *Donnachie* and *Landshoff* [13] which discuss, among other things, the problem of going from large to zero momentum-transfer. This, in other words, is the problem of calculating DIS from a bound state model.

References

1 C. Bernard et al.: Nucl. Phys. Suppl. **B30** (1992) 319

2 M. Schiffer, J. Bekenstein: Phys. Rev. **D39** (1989) 1109

3 "Mesons from Bag Model Compared to Lattice Theory and Heavy Ion Experiments", J. Dey, L. Tomio, M. Dey, S. Chakrabarty: IFT-UNESP preprint IFT P-022/93;" The Nucleon and Nuclei at Finite Temperature", M. Dey, S. Chakrabarty and J. Dey, IFT-UNESP preprint IFT P-035/93

4 S. Gupta: Nucl. Phys. Suppl. **B30** (1992) 351

5 M. Gonin: Nucl. Phys. **A553** (1993) 799c; F. Plasil, Opening Talk, International Conference on Physics and Astrophysics of Quark Gluon Plasma, Calcutta, India, January 1993

6 P. L. McGaughey et al.: Phys. Rev. Lett. **69** (1992) 1726

7 K. Gottfried: Phys. Rev. Lett. **18** (1967) 1154

8 J. Dey, M. Dey, L. Tomio, M. Schiffer: Phys. Lett. **A172** (1993) 203

9 NMC Collaboration, P. Amaudruz et al.: Phys. Lett. **B295** (1992) 66

10 A. E. Dorokhov, N. I. Kochelev: Phys. Lett. **B304** (1993) 167

11 F. E. Close: "The Nuclear Spin Crisis Bible", Rutherford Lab. preprint

12 E. Reya: "The Spin Structure of the Nucleon", Plenary talk presented at the Workshop on QCD - 20 Years Later (World Scientific, Singapore, 1993). Dortmund preprint DO-TH 92-17, p. 28

13 A. Donnachie, P. V. Landshoff: Manchester preprint no. M/C-TH 93/11, Cambridge preprint no. DAMTP 92-23

Subject Index

adjoint (see also representation) 38
algebra of anticommuting numbers 37-38
anomaly cancellation 2
anticommuting functions 38
antiscreening 81
antisymmetry of wave function 13
asymptotic freedom 83-84
axial rotation (see chiral rotation)
axial vector current 89

bag (model, see also MIT) 16-19
baryon current 84, 92
beauty baryons 71
Borel transform 88, 94

calorons 62
cancellation of spin orbit effects 22
canonical commutation relations 26
Casimir energy 16
charge fractionalization 30
charmed baryon 69-71
charmonium sector 89
chiral rotation (transformation) 4, 40
chiral symmetry 7
colour dielectric 81
compactification 48,
condensates
    gluon 81, 83, 89, 93
    quark 81, 83, 89, 93
conditionally convergent quantity 40
conformal transformation 57
correlators 34
conserved charge 25
constituent
    masses 3
    quarks 15
covariant derivative 74
CP non-invariance 54
current (algebra) 24-25, 72, 78
    commutators 26-27
    masses 3, 15, 67
    axial (see also, PCAC) 33

correlator 34.84 et seq.
cutoff 75

deep inelastic scattering 1
degenerate vacuum (see also multiple vacua)
    29
degrees of freedom 19
density 6-7, 23,
    dependence (see also high ... scaling
        with ...) 23
density of states 17-19
dibaryon 67
dilatation invariance 56
dimensional regularization 75,
dipole moment of the neutron 54
diquark binding (see also III) 70
Dirac-Maxwell 31
dual tensor 35,

external fields method 73

Fermi-Breit interaction 14, 67
finite action in Euclidean space 45
four fermion operator 87

gapless semiconductor 50
gauge field coupled to fermion [SU(N)] 39
Gell-Mann matrices 10, 66
Gell-Mann Okubo relation 68
GIM mechanism [Glashow, Iliopoulos and
    Maiani (1974)] 3
global transformation (chiral) 31-32
Goldstone boson (see also Nambu-...) 4, 72
Grassmann variables 76
GSI (particle) 28
gypsy (see also J/ψ) 1

heavy ion (HI) collisions 4-6, 28
hidden symmetry 20, 80
high density (see density) 6-7, 23
high temperature (T) 6-7
    shrinking of instanton size at high T 63

hypercharge   9, 25
hypersurface   47

incommensurability in charge density   50
instanton size   63
III, instanton induced interaction   67, 68
   third particle effective mass in III   68
   dependence on baryon (meson) size   70

J/ψ,   1, 4-5

large N (colour)   2, 20-23, 74
lattice calculations   7-8

Mandelstam variables   73, 84
mapping   47-49
   from spacetime to group manifold   49
Matsubara frequency   62
mean field approximation   21
mesonic mass spectrum   18-19
MIT bag (see also bag)   15, 67
multiple vacua   43

Nambu-Goldstone boson (see also Goldstone)
   3-4
non-Abelian guage theories   11-12, 36
nonperturbative effects   69
nonperturbative QCD fluctuation   43
nucleon mass (leading order)   94
nucleon propagator   76
nucleus-nucleus collisions (see heavy ion)

one gluon exchange (OGE)   14

parity doublets   7
partons   1
Pascual and Tarrach   84
PCAC   33, 90
Peccei Quinn Mechanism   55
Peierls and Yoccoz procedure   22
periodicity   61-62
permutation symmetry   13
pion-nucleon scattering amplitude Σ 77
planar gluon diagrams   21
polarization of the Dirac sea   37
polarized structure function of the proton   5
Pontryagin density   59
Pontryagin index   51, 57-59
pseudoscalar channels   90-91

Quantum Hall Effect   49
quark bilinear   84

representation
   adjoint   10
   fundamental   10-11
resonance 85 et seq.

scalar field   24
scalar operators (complete set for $d \leq 6$)   86
scalar superpotential   57
scale invariance   60
σ term   77
Skyrme model, Skyrmion   59-61, 79-81
source of the anomaly (non-Abelian)   40
specific heat   19
sphaleron   56
spin zero channel
   non-perturbative effect   91
strangeness   9, 12, 25
strange quark condensate   92
subtraction terms   88
SU(N)   2, 39, 52
$SU(3)_R \times SU(3)_L$ 72
SUSY   55
swelling of the nucleon   5

temperature (see also high T)   61
thermal mass   95
thermal pions   95
three point function   87
topological charge   30, 51
   current   29
   solution   30
   objects   49
   vacuum   53
transformations
   axial (see chiral)
   infinitesimal   11
tunnel   53
   tunneling   44-45, 65
two point function   87
two Skyrmion (configuration)   60

vacuum polarization 84 et seq.
   vector dominance   79-80

Ward identities   27-28, 35-37
Wilson coefficients   85, 95
winding number   48

zero energy (mode)   64
   configuration   46
   solution   30, 65

Citation Index

Actor (1979) 46
Adami, Hatsuda and Zahed (1991) 94
Adkins, Nappi and Witten (1983) 79
Adler (1969) 33
Affleck (1981) 56
Altarev, et al. (1986) 54
Atiyah Singer index theorem (1968) 36, 64
Atyiah, et al. (ADHM) (1978) 59
Atiyah and Ward (1977) 59
Atiyah and Manton (AM) (1989) 59

Balachandran et al. (1982, 83*) 79
Bando et al. (1988*) 20
Barut (1964, 68) 15
Belavin, Polyakov, Schwartz and Tyupkin (1975) 43
Bell and Jackiw (1969) 33
Belyaev and Ioffe (1983) 92, 94
Bernard (1979) 52
Bethe Salpeter (BS in short) (1951) 20
Bhaduri (1988) 14, 19, 44, 79–81
Bhaduri and Brack (1982) 15
Bhaduri, Cohler and Nogami (1980) 15
Bhaduri, Dey and Preston (1984) 16
Bhaduri, Dey and Srivastava (1985) 17
Bhaduri and Dey (1983) 19
Bhaduri et al. (1990) 80
Birse (1990) 81
Birulev et al. (1981) 76
Bjorken and Drell (1965) 39, 84
Bochkarev and Shaposhnikov (1986) 94
Bochkarev (1987) 94
Born et al. (1991) 7
Brown and Rho (1990, 91) 5
Brown and Rho (1989) 23
Brown et al. (1991) 7
Buchmüller and Wyler (1983) 55

Caldi (1977) 65, 91
Callan, Dashen and Gross (1978) 65
Capstick and Isgur (1986) 14
Carlitz and Cramer (1979) 65

Carlson (1985) 80
Cenni et al. (1985) 23
Cho et al. (1980) 76
Chodos et al. (1974) 15
Close (1979) 1, 12
Close, Roberts and Ross (1986) 5
Crater and van Alstine (1984) 20
Coleman and Witten (1980) 20
Coon and Scadron (Baryon 1992 reprint) 77
Cudell and Dienes (1992) 15, 17

De Tar and Kogut (1987) 7
Dey and Dey (1984) 15
Dey and Dey (1986) 5
Dey and Dey (1990) 15
Dey, Dey and Tomio (1992) 15, 17
Dey, Dey and Ghose (1989) 17
Dey et al. (1988) 5
Dey, Dey and Le Tourneux (1986) 20–22
Dey, Dey and Ghose (1985) 23, 94
Dey et al. (1990a) 23
Dey et al. (1990b) 23
Dey, Dey and Volkovitsky (1990) 67, 69–71
Dey, Dey and Ghose (1987) 80
Dey, Eletsky and Ioffe (1990) 95
Dittrich and Reuter (1985) 50
Donaldson et al. (1974) 76
Donoghue, Holstein and Klimpt (1984) 75
Dominguez and Loewe (1989, 91) 94
Dorokhov and Kochelev (1990a, 90b) 67, 94
Dosch and Narison (1988) 94

Eletsky and Ioffe (1988) 95
Engels et al. (1990) 7
Espriu et al. (1983) 92

Fai, Perry and Wilets (1988) 82
Forte (1989, 90) 67
Freund and Rosner (FR in short) (1992) 15, 17–19
Frederico et al. (1989) 23
Fritzsch (1990) 5

Fujikawa (1979, 80)   36-42
Fukugita et al. (1989, 90)   7
Furnstahl, Hatsuda and Lee (1990)   94

Gasser and Leutwyler (1982)   3, 68
Gasser and Leutwyler (1985)   74, 75
Gasser and Leutwyler (1984)   75
Gasser, Sainio E. and Svarc (1988)   76
Gasser et al. (1991)   77
Gavai, Potvin and Sanielevici (1989)   7
Gavai et al. (1990)   7
Gerber and Leutwyler (1985)   77
Gross and Wilczek (1973)   83
Gocksch, Rossi and Heller (1988)   7
Gottlieb et al. (1987)   7
Guichon (1988)   23
Gupta (1992)   8

Hagedorn and Rafelski (1980, 81)   17
Harrington and Shepard (1976)   62
Henley and Pashupathy (1992)   94
Hill et al. (1979)   76
Horn and Yankielowicz (1978)   67, 91
Huang and Stump (1976)   81

Isgur and Karl (1978, 79)   14, 67
Isgur et al. (1982)   15
Itzykson and Zuber (1980)   3, 27, 56
Ioffe and Karliner (1990)   67
Ioffe (1981)   92, 94

Jackiw (1985)   27, 56
Jackiw and Rebbi (1978)   30
Jackiw (1984)   30
Jackiw and Rajaraman (1985)   33
Jackiw, Nohl and Rebbi (JNR) (1977)   57
Jackiw and Rebbi (1976)   58
Jackson and Rho (1983)   79
Jaffe and Lipkin (1991)   6
Jain et al. (1989)   80
Jamin (1988)   88
Jennings and Bhaduri (1982)   17

Kalman and Tran (1989)   14
Kapusta (1981, 82)   17
Khatsimovskii et al. (1987)   55
Klinkhamer and Manton (1984)   56
Koch et al. (1992)   7
Kochelev et al. (1985)   67
Kogut et al. (1991)   7
Krein et al. (1991)   82
Kutasov and Seiberg (1991)   15, 17
Kuzmin, Rubakov and Shaposhnikov (1985)   56

Lacombe et al. (1988)   21
Leal Ferreira (1977)   19
Leal Ferreira and Zagury (1977)   19
Leal Ferreira et al. (1980)   19
Leal Ferreira (1989)   21
Lee (1981)   54, 81
Leech and Birse (1992)   82
Leutwyler and Roos (1984)   75
Lucha et al. (1990)   14, 20

Matsui and Satz (1989)   1
McLerran, Vainshtein and Voloshin (1990)   56
McLerran (1986)   61
Mignaco and Wulck (1989)   80
Milton (1983)   16
Mitra and Rajaraman (1989)   34
Murthy et al. (1984)   14
Murthy and Bhaduri (1985)   15
Muta (1987)   1

Narison (1989)   84
Nagamiya (1988)   5
Nielsen and Olesen (1973)   46
Nielsen and Ninomiya (1983)   50
Nilles (1984)   55
Novikov (1981)   91
Novozhilov (1990)   20, 79

Oka and Takeuchi (1991)   66
Ovchinnikov et al. (1988)   94

Palladino and Leal Ferreira (1989)   19
Pascual and Tarrach   84
Pokorski (1987)   2, 56
Polchinski and Wise (1983)   55
Politzer (1973)   83
Preparata et al. (1990)   67

Rabinowitz et al. (1992)   67
Rajaraman (1982)   38, 47, 59
Reinders, Rubinstein and Yazaki (1982, 85)   23, 84
Richardson (two body potential) (1979)   20-23
Ringwald (1990)   56
de Rujula, Georgi and Glashow (1975)   14, 67-68

Satz (1988) (see also Matsui)   5
Schwinger et al. (1979)   16
Schwinger terms   27
Schwinger (1951)   28
Schrieffer (1985)   30

Semenoff and Sodano (1986)   49
Sen (1991)   57
Shifman et al. (1980)   65
Shifman, Vainshtein and Zakharov (SVZ) (1979)
    83
Shuryak (1988)   23, 65, 91
Shuryak (1988a)   63
Shuryak (1988c)   65
Shuryak and Verbaarschot (1991)   65
Shuryak and Rosner (1989)   67–70
Shuryak (1983)   67, 91
Shuryak (1982)   91
Silvestre-Brac and Gignoux (1985)   14
Skyrme (1961)   20, 28
Smith et al. (1990)   54
Su, Schrieffer and Heeger (1979, 80)   30
Su and Sakita (1986)   50
Suzuki et al. (1985)   16

Takeuchi and Oka (1991)   67
Tegen et al. (1982)   20
't Hooft (1974)   20
't Hooft (1976b)   64
't Hooft (1976a)   65
Toki, Dey and Dey (1983)   15

Ukawa et al. (1989)   7

Wilets (1989)   81
Wilson (1969)   85
Witten (1979)   20, 21

Zahed and Brown (1986)   59, 79
Zu-Rong Yu (1988)   15

Springer

Springer-Verlag
and the Environment

We at Springer-Verlag firmly believe that an international science publisher has a special obligation to the environment, and our corporate policies consistently reflect this conviction.

We also expect our business partners – paper mills, printers, packaging manufacturers, etc. – to commit themselves to using environmentally friendly materials and production processes.

The paper in this book is made from low- or no-chlorine pulp and is acid free, in conformance with international standards for paper permanency.